p86-97
120-126

Rehabilitation, Crime and Justice

Rehabilitation, Crime and Justice

Peter Raynor

and

Gwen Robinson

Consultant Editor
Jo Campling

First published 2005 by
PALGRAVE MACMILLAN
Houndmills, Basingstoke, Hampshire RG21 6XS and
175 Fifth Avenue, New York, N.Y. 10010
Companies and representatives throughout the world

PALGRAVE MACMILLAN is the global academic imprint of the Palgrave Macmillan division of St. Martin's Press, LLC and of Palgrave Macmillan Ltd. Macmillan® is a registered trademark in the United States, United Kingdom and other countries. Palgrave is a registered trademark in the European Union and other countries.

ISBN-13: 978–0–333–68740–6 hardback
ISBN-10: 0–333–68740–X hardback

This book is printed on paper suitable for recycling and made from fully managed and sustained forest sources.

A catalogue record for this book is available from the British Library.

Library of Congress Cataloging-in-Publication Data
Raynor, Peter.
    Rehabilitation, crime and justice / Peter Raynor and Gwen Robinson.
        p.   cm.
    Includes bibliographical references and index.
    ISBN 0–333–68740–X (cloth)
    1. Criminals—Rehabilitation.   2. Criminal justice, Administration of.
    3. Criminal psychology.   I. Robinson, Gwen, 1969–   II. Title.
    HV9275.R38 2005
    365′.66—dc22   2005047532

10   9   8   7   6   5   4   3   2   1
14   13   12   11   10   09   08   07   06   05

Printed and bound in Great Britain by
Antony Rowe Ltd, Chippenham and Eastbourne

Contents

Acknowledgements   vi

1 Defining Rehabilitation   1
2 Justifying Rehabilitation   16
3 Origins and Contexts   32
4 The Rehabilitative Ideal: Advance and Temporary Retreat   49
5 Adapting to the End of 'Treatment'   78
6 The New Rehabilitation: 'What Works' and Corrections
   at the End of the Twentieth Century   98
7 Against the Tide: Non-treatment Paradigms for the
   Twenty-first Century   134
8 The Futures of Rehabilitation   159

Notes   173
References   179
Index   201

Acknowledgements

The authors would like to thank their respective partners, Jennifer and Martin, for putting up with the process of producing this book. In addition, Peter is grateful to the Nuffield Foundation for the Social Science Research Fellowship which supported the beginning of this work as long ago as 1994, and Gwen is grateful to the University of Wales, Swansea for the award of a studentship to support her PhD studies from 1998 to 2000. Gwen would also like to express her thanks to colleagues at the University of Sheffield and elsewhere who have encouraged and supported her work. Peter would also like to thank Brian Willan for permission to use some material which originally appeared in *Alternatives to Prison* (edited by Bottoms, Rex and Robinson).

1
Defining Rehabilitation

> No idea is more pervaded with ambiguity than the notion of reform or rehabilitation.
>
> – Allen (1959), p. 226

As recently as the early 1980s, academic commentators were writing about rehabilitation as an approach towards offenders that had had its day. In the wake of the well-documented collapse of the 'rehabilitative ideal' (Allen, 1959), critics and supporters alike seemed to converge on the idea that rehabilitation as a penal strategy would soon be a thing of the past (e.g. Bottomley, 1980; Bottoms, 1980a; Allen, 1981). This conclusion was powerfully re-stated a decade later by American criminologists Feeley and Simon (1992, 1994), whose 'new penology' thesis seemed to confirm the death of rehabilitation in Western penal systems.

However, less than a decade later, rehabilitation was beginning to enjoy a marked revival. By the late 1990s, we were witnessing renewed commitment to rehabilitation as an approach towards offenders. This revival was not confined to the realm of penal practices, but was also evident in (and legitimated by) penal policy statements, as well as a significant injection of resources for the development and evaluation of rehabilitative programmes for offenders. For example, in April 1999, the UK Government launched a wide-ranging initiative intended to develop evidence-based strategies for reducing crime. This Crime Reduction Programme included in excess of £20 million to support the development of a range of programmes aiming to reduce recidivism among known offenders (Maguire, 2004a; Raynor, 2004a).

In parallel with this somewhat surprising trend, much has been written about rehabilitation in the last decade or so. However, there has been relatively little consideration of just what rehabilitation means in the context of offending. Whereas we might have expected the revival of rehabilitation to be accompanied by attempts to clarify or refine the term, what we have seen instead is a proliferation of *new* terms and concepts allied with rehabilitation, many of which are far from transparent, for example 'new rehabilitationism' (Hudson, 1987) and 'born-again' rehabilitation (O'Malley, 2001). We thus have a whole new vocabulary around rehabilitation, but we are arguably no closer to a common understanding of just what the rehabilitation of offenders entails. Indeed, it is arguably the case that we have exacerbated the confusion expressed by Francis Allen almost half a century ago.

What this means is that, in our discussions of offender rehabilitation – whether in academic, policy or practice contexts – it is far from clear whether we are all speaking the same language. In short, we cannot be sure whether those who discuss and/or promote offender rehabilitation share a common vision of the 'rehabilitative enterprise'. This is an important point, and more than a matter of semantics. If we cannot say what rehabilitation is, how can we begin to account for its revival? Moreover, how can we judge whether it is effective, or just, or humane?

The primary aim of this introductory chapter is to provide a conceptual context for the analysis of offender rehabilitation, past, present and future, which is the major focus of this book. Our aim in this chapter is not to arrive at a single definition of rehabilitation, but rather to consider a range of ways in which the rehabilitation of offenders has been conceived, thought about and understood. In essence, we aim to *problematise* the concept which is at the heart of this book. To this end we take a deliberately broad-brush approach, inviting consideration of a variety of 'ways of seeing' this important concept. We begin by considering some generic or non-criminological definitions of rehabilitation, before going on to present a typology of perspectives on offender rehabilitation. We then reflect back on the degree of convergence between criminological and non-criminological understandings of this key term. The chapter concludes with an outline of the structure and content of the book.

Back to basics: Generic and medical definitions

Let us begin our conceptual exploration by considering a generic, dictionary definition. The Oxford English Dictionary (OED) defines rehabilitation as: 'the action of restoring something to a previous (proper) condition or status'. What can we infer about rehabilitation from this brief definition? First, it is clear that rehabilitation is understood as an *action* or *process*. Secondly, this process is one which is focused on the notion of *restoration*, which seems to be at once regressive and progressive. On the one hand, restoration involves a return to some former state (regression); but on the other hand this state is considered the proper or preferable state, the implication being that it is a progressive or positive move: a change 'for the better'. Another way of putting this would be to say that the rehabilitation process follows from or 'corrects' some process of deterioration or recession. A final observation is that the 'object' of rehabilitation is not specified: we can infer that it is a process which may be applied or applicable to a range of objects. Reflecting this it is worth observing that we do tend to talk about the rehabilitation – implying 'revival' – of ideas or fashions, as well as people.

Moving on to consider how rehabilitation might apply to individuals, the OED helpfully supplies a supplementary definition. This refers to: 'restoration of a disabled person, a criminal etc., to some degree of normal life by appropriate training etc'. In this definition, the notion of *restoration* is again central: there is a clear sense that rehabilitation involves getting 'back to normal' and that it may thus be applied to any person who has strayed from a 'norm' of some kind. It is worth noting in this context that in everyday parlance, rehabilitation is commonly associated with the abuse of alcohol or illegal drugs, where it implies the achievement of abstinence or at least use within 'normal' limits. We also see, in the notion of 'appropriate training', the introduction of the idea of *intervention*: that is, the presence of a third party is implied, the role of whom is to assist in the process of rehabilitation. This definition also specifies the typical groups of people for whom we tend to consider the notion of rehabilitation relevant, namely 'disabled persons' and 'criminals'. We will come on to 'criminals' or offenders (our preferred term) in a moment, but first let us consider a further, non-criminological definition of rehabilitation, namely one derived from medicine.

According to a medical definition, rehabilitation denotes: 'the restoration to health and working capacity of a person incapacitated by disease, mental or physical, or by injury' (Macpherson, 1992). In this context, rehabilitation is clearly understood as a process which follows a mental or physical setback or deterioration: the subject has become ill or sustained an injury, and rehabilitation clearly refers to efforts to re-establish the subject's former, 'healthy' status of physical or mental fitness or well-being. This shares much in common with the generic definition considered above: that is, rehabilitation is understood as a process of getting 'back to normal'. In this definition there is no reference to 'appropriate training', but for most of us the notion of rehabilitating the sick or injured will conjure up images of physical and occupational therapists and other medical experts, all intervening in appropriate ways to enable the person's recovery.

The rehabilitation of offenders: Criminological definitions

Having considered some non-criminological definitions of rehabilitation, let us now move on to the subject of this book, namely the meaning(s) of rehabilitation in the context of offending. We have already noted that, despite the longevity and continuing relevance of the concept of rehabilitation in the context of offending, it has rarely been 'unpacked' or examined critically. Indeed, it is quite common to come across 'offender rehabilitation' in both academic and policy contexts with no accompanying definition of the term. Even where rehabilitation has been defined, we have not always been left with a crystal clear picture. Consider, for example, the following definition, offered by Hudson:

> taking away the desire to offend, is the aim of reformist or rehabili-
> tative punishment. The objective of reform or rehabilitation is to
> reintegrate the offender into society after a period of punishment,
> and to design the content of the punishment so as to achieve this.
> (Hudson, 2003, p. 26)

This is a relatively brief definition; but it is far from straightforward. Indeed, it throws up a number of questions. For one thing, Hudson's definition is unclear about what the objective of rehabilitation is. Is it taking away the desire to offend, or is it 'reintegration into society'? Are these objectives compatible, or quite different from each other? This definition also leaves us wondering about the relationship(s) between rehabilitation and punishment. On the one hand, Hudson seems to be saying that rehabilitation is a form of punishment; but on the other hand she implies that it follows a period of punishment. Another potential source of confusion is Hudson's 'coupling' of notions of reform and rehabilitation. Again, we are left wondering whether these concepts are interchangeable, or whether they denote similar but subtly different processes.

Offender rehabilitation: Developing a typology

In this section we move away from existing definitions and attempt instead to develop a broader range of understandings: that is, a typology of perspectives on offender rehabilitation. After considering these perspectives we will refer back to the non-criminological definitions explored earlier and consider the extent to which these are reflected in our own typology.

'Correctional' rehabilitation

The first model of rehabilitation we shall consider views rehabilitation as a process of effecting positive change in individuals. We will refer to this as the 'correctional' model of rehabilitation. It is regarded by many as the 'orthodox' model of rehabilitation, and it is the model most closely associated with notions of *treatment* in the context of offending (e.g. Hollin, 2001; McGuire, 2002). In other words, the model which we outline in this section incorporates what is commonly referred to as the *treatment model*. Correctional rehabilitation can also, however, refer to a training or learning-centred model, in which the acquisition of skills is central.

At the heart of the correctional model is the notion of *corrigibility*: that is, a belief in the propensity of (at least some) offenders to change 'for the better'. The correctional model is often associated with the objective of desistance, or the cessation of offending; however, it is sometimes associated with a broader aim which involves more fundamental change. The correctional model assumes that positive change, however conceived, can be brought about by subjecting offenders to particular interventions, programmes or regimes: with the right intervention, offenders can be brought into line with a (law-abiding) 'norm'. The idea of correction is therefore closely associated with the idea of 'normalisation': that is, the adjustment of a person's character or behaviour towards some kind of a socially acceptable model (Foucault, 1977; Garland, 1985). In Foucault's work, 'normalisation' is the objective of 'disciplinary' interventions.

Although it comprises a coherent model of rehabilitation for the purposes of our typology, the correctional model is in fact a very broad one, in that it encompasses a wide variety of theories and approaches. The source of the internal diversity of the correctional model is the fact that it is dependent on the generation of theories about why people offend. These theories then suggest targets for and methods of intervention. So, whilst the correctional model is intimately bound up with theories about the causes of offending, it does not subscribe to any particular theory or body of theories. In other words, the correctional model supports a variety of 'criminologies', as well as a variety of methods or interventions.

This aspect of the correctional model is nicely captured in Hollin's definition of rehabilitation:

> The rehabilitation model takes the stance that crime is best prevented by addressing directly the factors – economic, social, or personal – believed to be the cause of crime. (2001, p. 242)

Hollin clearly understands rehabilitation as a process of tackling the causes of crime; but his definition potentially supports a number of theories – economic, social and personal – about the causes of crime and thus the appropriate 'targets' for intervention.

Notwithstanding this theoretical flexibility, it is fair to say that the correctional model is *principally* associated with criminologies which locate the causes of offending in individual offenders, rather than in external factors. As such, it exemplifies 'the strong and persistent inclination to assign the cause of crime to the endogenous constitution of the criminal' (Radzinowicz and Hood, 1990, p. 3). Reflecting these assumptions about the causes of offending, it is principally concerned with effecting *change in offenders themselves*, rather than in their social, economic or physical situation (the latter may be seen as desirable, but more difficult to achieve, or requiring large-scale social action beyond the scope of 'correctional' agencies). It assumes that it is possible to isolate or identify the causes of offending – whether they are related to the offender's character, morality, personality, psychological make-up or choices – and then intervene in ways which will remove those causes. In short, then, the correctional model of rehabilitation seeks to *remove or 'undo' the causes of offending* (Garland, 1985).

The correctional model is very much associated with notions of 'rehabilitative punishment' (Hudson, 2003) or 'penal rehabilitationism' (von Hirsch and Maher, 1992). This is not, however, to say that the correctional model is *synonymous with* punishment: rather, that punishment is one context in which the correctional model of rehabilitation can be found. In other words, when rehabilitation has been linked with punishment, it has tended to be a correctional model of rehabilitation which is implied.

Rehabilitation and reform

As we have acknowledged, the correctional model is to a large extent a theoretical construct; an 'umbrella term' used by us to denote a particular 'way of seeing' the rehabilitative enterprise. It should be noted that, as a theoretical construct, the correctional model eschews the distinction between *rehabilitation* and *reform* which is drawn by a number of penal historians (e.g. Bottoms, 1980a; Garland, 1985; Hudson, 2003). It will be recalled that in Hudson's definition, considered above, 'reformative' and 'rehabilitative' interventions were referred to as a singular enterprise. This coupling of reform and rehabilitation is not uncommon, and it reflects a view of both reform and rehabilitation as 'correctional' approaches. However, rehabilitation is often theorised as an approach distinct from reform: that is, as a *particular style* of correctional intervention and a product or correlate of a particular historical context.

As Hudson has explained,

> My own preference is to use 'reform' for the nineteenth-century development of regimes designed to effect change in individuals through educative and contemplative techniques, and to use 'rehabilitation' to signify the more individualistic treatment programmes that became established during the twentieth century. (Hudson, 2003, p. 27)

This quotation indicates, first, that reform and rehabilitation have much in common: both appear to refer to efforts to bring about positive change in individuals; but it is also clear that Hudson is describing two contrasting models of 'correction'. An historical contrast is implied: reform is a nineteenth-century development whereas rehabilitation 'became established in the twentieth century'. A further contrast is the methods employed by reform and rehabilitation: Hudson equates 'reform' with 'educative and contemplative techniques', and 'rehabilitation' with 'treatment programmes'. The latter are also characterised as 'more individualistic', which implies the tailoring of 'treatment' to individuals, whilst also indicating that reform is associated with a more blanket approach. It is this move towards the individualisation of intervention which, for some historians and sociologists, characterises the shift from reform to rehabilitation (e.g. Garland, 1985) – or, in the narrower context of penal sanctions, from reformative to rehabilitative punishment (Hudson, 2003).

Beyond the correctional model: Reintegration and resettlement

We noted that, for many, the correctional model constitutes the 'orthodox' model of rehabilitation. However, we also noted that it by no means constitutes the *only* way of understanding the rehabilitation of offenders.

If we return for a moment to Hudson's definition, it will be remembered that reference was made to attempts 'to reintegrate the offender into society'. This 'reading' of rehabilitation is in fact a relatively common one, and is not always understood as being at odds with 'correctional' intervention. For example, Crow (2001) views rehabilitation as a necessary complement to 'correctional' treatment programmes. Without rehabilitation, Crow argues, the treatment process is at best incomplete, and at worst runs the risk of failure.

How, then, does Crow define rehabilitation? Crow uses the term to refer to 'a wide variety of interventions directed towards the *social reintegration* of offenders [including] schemes which aim to provide offenders with accommodation, education, training and employment' (2001, p. 8; emphasis in original). Rehabilitation, he argues, is about change, but not in the 'correctional' sense. With reference to a dictionary definition, Crow explains that he understands rehabilitation as being concerned with change that is *restorative*: that is, which seeks to return the offender 'to a position in society which they formally held' (2001, p. 4). In contemporary discourse, we can equate this model of rehabilitation with attempts to promote *social inclusion* (Pease, 1999).

Crow is one of a number of writers who have linked notions of rehabilitation and *resettlement*. 'Resettlement' is a term usually used to refer to a variety of interventions, the common objective of which is the reintegration of the offender into society after a period of imprisonment. This notion of resettlement resembles quite closely the medical definition of rehabilitation discussed earlier. It will be recalled that this definition referred to the restoration of a person *after a period of incapacitation*. In the medical context, incapacitation refers to inactivity and often refers to a period spent in hospital, but in the context of offending, incapacitation often refers to a period of imprisonment. Just as illness or disability may be understood as incapacitating, so too might a period spent in custody.

Although understandings of rehabilitation as social inclusion or reintegration do not necessitate specific criminologies, or theories about why people offend, they do tend to favour *sociological* theories about crime causation. That is, they tend to emphasise the social and/or economic causes or correlates with offending. As such they are sometimes understood as representing a departure from the correctional model of rehabilitation which, it will be recalled, tends to emphasise the endogenous (internal/personal/individual) rather than exogenous causes of crime.

Another characteristic of this model of rehabilitation is that it tends to understand rehabilitation as a process which occurs independently of (though often after) a period of punishment. In other words, rehabilitation and punishment are conceptually divorced.[1] Indeed, for some, rehabilitation is understood as a process of *compensating for* or *undoing* any harmful effects of punishment. Here, rehabilitation is viewed in the context of punishment, but not as part of the punishment process: punishment is not understood as a vehicle for rehabilitation. This is arguably the least discussed model of offender rehabilitation; nevertheless, it is an important interpretation and merits mention. Far from seeing rehabilitation as part of a positive process of punishment, this model tends to view rehabilitation as an *antidote to* punishment, or a means of at least limiting punishment in accordance with principles of desert or proportionality. This model concedes that punishment may be deserved, but argues that just as the state has a right to punish the offender, so the offender has a corresponding right not to be unduly disadvantaged by the experience of punishment. Any handicaps or damage created by punishment ought to be offset by rehabilitative measures (e.g. Cullen and Gilbert, 1982; McWilliams and Pease, 1990).

In the following section we introduce the legal notion of rehabilitation, which has much to say about rehabilitation as a process of reintegration or social inclusion. Arguably one of the most surprising aspects of the academic literature on offender rehabilitation is the tendency to omit any discussion of rehabilitation in the legal context. However, in our view this is a crucial model of rehabilitation, not least because it probably has the longest history of all the models we have considered. It is also valuable in that it offers a new perspective on the notion of rehabilitation as 'restoration', and it brings into question the necessity of intervention in our understanding of offender rehabilitation.

Rehabilitation and the law

We saw that rehabilitation has been understood as an endeavour to remove the desire to offend, or as undoing the causes of offending. However, as Garland (1985) has pointed out, this is not in fact the original meaning of the term. According to Mannheim (1939), the concept of rehabilitation was first conceived in French law in the second half of the seventeenth century and, in this context, it referred to the destruction or 'undoing' of a criminal conviction: 'the effect of an act of rehabilitation usually consists in a deletion of all entries regarding the conviction in the records' (p. 151).

What is the purpose of deleting a criminal conviction? Let us look in a bit more detail into the legal model of rehabilitation in UK law, as enshrined in the 1974 Rehabilitation of Offenders Act. This Act was passed largely in response to the recommendations of a committee set up jointly in the early 1970s by pressure groups JUSTICE, the Howard League for Penal Reform and the National Association for the Care and Resettlement of Offenders. The committee was set up to consider the problems of a criminal record to 'rehabilitated persons', defined as the large number of people 'who offend once, or a few times, pay the penalty which the courts impose on them, and then settle down to become hard-working and respectable citizens' (JUSTICE *et al.*, 1972, p. 5). Noting that the United Kingdom was at that time the only member country of the Council of Europe with no law relating to rehabilitated persons, the committee made a strong case for 'rehabilitation laws' which would treat the majority of old convictions as 'spent and irrelevant' (p. 13).

That the reintegration of offenders into society was central to the deliberations of the committee is evident in the title of its report: *Living it Down: The Problem of Old Convictions*. The report argued that rehabilitated persons commonly faced discrimination in the fields of employment, insurance, and in the courts, because their criminal record continued to mark them out as offenders well after they had turned their backs on crime. As the report's authors explained:

> At any time, on any day, malice or chance may put an end to [a person's] rehabilitation, and expose them to endless unemployment and misery [. . .] For rehabilitation to be complete, society [. . .] has to accept that [rehabilitated persons] *are* now respectable citizens, and no longer to hold their past against them. (JUSTICE *et al.*, 1972, pp. 3–5)

It is clear from this statement that the legal notion of rehabilitation sees the act of physically deleting a person's criminal record as performing an important symbolic function: that is, it serves to 'de-label' a person, enabling him or her to shed a negative (criminal) identity and (re-) assume a positive, non-criminal one. Reflecting the report's title, then, the legal notion of rehabilitation is very much about 'living down' the stigma of a conviction and reducing societal discrimination against that person: it is about reinstating the person as a law-abiding citizen. This view is also expressed by Faulkner, as follows: 'No offender should be denied a public service or facility to which they would ordinarily be entitled, or expected to accept a lower standard, simply because they have a conviction' (2003, p. 307).

Here we see a model of rehabilitation founded on the notion of restoration which, it will be remembered, was so central to our generic definitions. According to the legal model of rehabilitation, *restoration of the state of citizenship* is its clear, principal objective. This process of restoration involves adopting the rights and obligations of 'normal' citizens: the rehabilitated person is perceived as having paid the penalty for his or her offence(s) and as subsequently showing a commitment to 'going straight'. As such, the legal model views the rehabilitated person as deserving of a return to the community of law-abiding citizens and restoration to the status and entitlements of full citizenship.

Comparing generic and criminological models of rehabilitation

Let us return for a moment to the non-criminological definitions of rehabilitation which we considered at the beginning of this chapter. We learned that in these contexts, rehabilitation tends to denote four things. First, it implies a process. Secondly, it is understood as a positive or progressive process: a 'change for the better'. Thirdly, it implies restoration, which essentially means returning to some former state. Fourthly, it usually implies the intervention of some other party.

How well do these elements describe rehabilitation in the context of offending? In general terms we can agree that, in the context of offending, rehabilitation is generally understood as a process, albeit that the nature of that process is not one on which all are agreed. It is also generally understood as a process which is essentially positive: that is, offender rehabilitation is usually considered to be a good thing, not just for offenders themselves but also, frequently, for the communities in which they live.

What about rehabilitation as restoration? Clearly it is *possible* to understand offender rehabilitation in terms of a process of restoration. One model which has restoration at its heart is the legal model, which understands rehabilitation as a process whereby the offender re-acquires his or her standing as a citizen. Further, we saw in the context of the 'social re-integration' model that Crow (2001) refers to rehabilitation as a restorative process. However, the notion of restoration is not entirely straightforward in the context of offending and terms like 'rehabilitation', 'resettlement' and 'reintegration' are all potentially problematic on the grounds that they imply restoration of a condition that possibly never existed. As has recently been pointed out, 'many imprisoned offenders were not habilitated, integrated or settled prior to their incarceration' (Home Office, 2001, p. 3). For many offenders, then, restoration to a former state or situation is not necessarily a feasible or desirable objective.

It is worth noting, then, that not all of the models considered above view restoration as a process of putting the offender back in his or her former position. For example, in the correctional model there is no assumption that the offender has ever conformed to the desired norm: the norm is a societal standard rather than one relating to the offender's previous standing. Reflecting on the generic definition, then, rehabilitation may involve restoration to *either* a previous *or* a 'proper' condition/ status, although how the latter is defined will depend on the perspective adopted.

Another way of thinking about restoration in the context of offender rehabilitation is in terms of a process of 'undoing'. In the correctional model, rehabilitation involves the 'undoing' of the causes of offending, and the legal model seeks to physically 'undo' a conviction. It is through these various processes of 'undoing' that the offender is arguably thought to be restored – whether to the 'norm' of law-abiding citizen, or to his or her former legal status.

Finally, intervention. We saw in the non-penal definitions discussed at the beginning of this chapter that the process of rehabilitation is often associated with the intervention of a third party (or parties). For example, the rehabilitation of a person affected by injury is likely to involve the intervention of medical and other experts to assist in the process of recovery. It is clear that intervention is central to some of the penal definitions we have considered. This is particularly true of the correctional model. As Crow (2001) has pointed out, the notion of correction or treatment is often regarded as something that is 'done to' someone, usually by a person in a position of expertise, authority and a certain degree of power. 'Rehabilitative punishment' certainly implies the involvement of the state, or else some state-sanctioned authority, in the design and/or delivery of the appropriate intervention. Indeed, rehabilitation in this context can be understood as an exercise of state power (e.g. Cohen, 1985).

However, in other models of rehabilitation, intervention is not considered a necessary ingredient. For example, the legal model does not assume intervention in the process of rehabilitation. Although rehabilitation is clearly linked with desistance, or the cessation of offending, the legal model does not concern itself with *bringing about* desistance in the way that the correctional model does. The rehabilitated person is someone who has desisted from crime for a specified period; but there is no assumption that he or she has been assisted or 'helped' in this process by any third party. The report by JUSTICE *et al.* (1972) contains no reference to expert intervention in its discussions of rehabilitated persons. Rather, rehabilitation is represented as a 'natural' process that applies to the majority of those who are convicted of a criminal offence and several references are made to the notion of persons having reportedly *rehabilitated themselves*. Those who commit one or a small number of offences in adolescence are thought simply to 'grow out of the need to behave delinquently' (1972, p. 5). We will return in Chapter 7 to this notion of 'growing out of crime', but for the moment we need only note that the legal model introduces the possibility of 'do-it-yourself' rehabilitation, which is a departure from the orthodox understanding – though it is presumably what the majority of offenders, who receive 'non-rehabilitative' penalties, are expected to achieve. Clearly then the legal model raises questions about whether the offender ought to be understood as a passive recipient of 'help' or 'treatment', or an active participant in his or her own rehabilitation.

Rehabilitation and eligibility

Before concluding this introductory chapter, we pose one final question: to what extent is rehabilitation regarded as a universal notion in the context of offending? Put another way, who is eligible? Once again, the models we have discussed take different views on this issue. If we consider rehabilitative punishment, then straight away it is clear that not all offenders are regarded as eligible for such sanctions, however defined. Not all punishments can be considered rehabilitative and the choice of punishments may be determined by offence rather than offender factors. Further, eligibility is often linked with perceptions or assessments of 'corrigibility'. For example, the correctional model has to ask itself whether the offender is likely to respond to correctional interventions: is he or she amenable to attempts to bring about change?

The legal model also sets some limits to eligibility: that is, it does not regard all offenders in equal terms. For one thing, the Rehabilitation of Offenders Act (1974) does not apply to all sentences. For example, prison sentences of more than two-and-a-half years never become 'spent'. There are also some offices and occupations in which individuals are expected to declare their convictions, even if they are spent, under the Act. These include professions in areas such as health, pharmacy and the law; senior managers in banking and financial services and occupations which involve working with children and/or vulnerable adults (e.g. social work). The Act also contains provisions so that a subsequent conviction can extend or even invalidate the rehabilitation period for an earlier conviction.

Rehabilitation, crime and justice: Questions and issues

What has been the purpose of the foregoing analysis of offender rehabilitation? First of all, it alerts us to the fact that there appears to be no universally agreed definition of rehabilitation in the context of offending. Far from being a uniform concept, the rehabilitation of offenders appears to refer to a variety of processes and outcomes, not all of which are entirely compatible. This means that we cannot necessarily assume that those who contribute to discussions about the merits and problems associated with offender rehabilitation share the same understanding or vision of the 'rehabilitative enterprise'. A related point is that any assessment of offender rehabilitation is inherently problematic, unless the precise model of rehabilitation is defined. Having problematised the concept of offender rehabilitation, then, we have also in effect problematised most of the questions which have traditionally been posed in respect of offender rehabilitation, such as whether it is effective, humane or just (von Hirsch and Maher, 1992). We have not, however, *invalidated* these questions. Rather, we have argued for the importance of specificity in our discussions – and evaluations – of offender rehabilitation.

Outline of the book

This book aims to address these questions through an exploration of the history, effectiveness and prospects of rehabilitation. First, however, it follows from the arguments in this chapter that not only are the nature and meaning of rehabilitation problematic; so too are its qualifications. Chapter 2 considers some of the justifications traditionally advanced for rehabilitation, and their relationship to different definitions and aims.

From there, we proceed to an analysis of the theoretical, policy and practice histories of rehabilitation. In Chapter 3 we examine the emergence of a 'reformative ideal', focusing in particular on the period between the late eighteenth and the late nineteenth centuries, when ideas began to emerge about how the prison might be made to work in the interests of offender rehabilitation. The historical ground covered in this chapter takes us up the publication of the 1895 Gladstone Report, which is commonly viewed as a key milestone in the history of rehabilitation.

Chapter 4 continues this historical journey by considering how, during the middle decades of the twentieth century, rehabilitation became largely identified with the aims and methods of the emerging profession of social work and the so-called 'treatment model'. It goes on to describe how rehabilitation subsequently suffered a major setback, when a modernist optimism about the efficacy of a claimed scientific and professional approach gave way to a late modern doctrine that 'nothing works'.

In Chapter 5 we consider how penal practice adapted to the demise of the treatment model when, in the 1980s and 1990s, penal policy became increasingly preoccupied with the ostensibly non-rehabilitative concerns of decarceration and, subsequently, public protection. In this chapter we argue that despite sparse references to rehabilitation in policy discourse, rehabilitative impulses essentially adapted and survived until the emergence of new evidence about the effectiveness of 'correctional' practices served to renew the credibility and legitimacy of rehabilitation in the last decade of the twentieth century. It is the emergence and fate of this 'new rehabilitationism', more commonly referred to as the 'what works' movement, which concerns us in Chapter 6.

Chapter 7 moves away from the more explicitly 'correctional' modes of intervention, turning to what we characterise as a number of 'non-treatment paradigms': that is, contemporary developments in theory and practice which do not fit within the correctional mould, but which nonetheless clearly belong under the umbrella of rehabilitative initiatives, broadly defined. In this chapter we consider the contribution of restorative justice; developments in theory and research around desistance or 'naturalistic' rehabilitation; and the resettlement or reintegration of ex-prisoners. We conclude this chapter with a summary of what we think these approaches share in common, and what they have to offer those engaged in rehabilitative work with offenders.

Finally, in Chapter 8, we ponder the likely futures of rehabilitation. We ask whether, in the contemporary political climate, rehabilitation is doomed to extinction; or whether it is possible to envisage a brighter future for the ideals which have informed its development, and for the contemporary practices which we identify as promising. On the whole, we conclude, our survey of the history and the evidence gives us grounds for a guarded optimism.

2
Justifying Rehabilitation

As the previous chapter has shown, the rehabilitation of offenders can be defined and understood in a number of different ways. At different points in the history of modern penal systems different models of rehabilitation have been current, and each of them has different implications for policy, for sentencing and for direct practice with offenders. Each model also carries with it, in explicit or implicit forms, a set of arguments about why it is worth doing. An activity which is complex, expensive, difficult and unsure of success needs arguments in its support when it competes for resources, or needs to establish its claims against other aims of sentencing such as deterrence or incapacitation. In short, the desirability of rehabilitation is not always taken for granted, and its advocates need from time to time to deploy defences or justification of what they propose. The focus of this chapter is on the different kinds of justifications that have been offered to support various models of rehabilitation. It also considers the kinds of evidence on which they have drawn, and the assumptions they make about the nature, characteristics and entitlements of people who commit offences.

At the outset it is important to realize that this is not simply a matter of deploying evidence about effectiveness. In recent years, as subsequent chapters will show in detail, we have tended to take for granted that the most important question to ask about rehabilitation is 'What works?' or, in other words, what means can we use to pursue our goals in the most effective manner. There has been less discussion about what those goals should actually be. Usually the implied goal seems to be less re-offending by sentenced offenders, but other versions of rehabilitation have pursued very different goals: for example, the salvation of human souls, or the healing of damaged relationships, or a greater sense of safety or security in the everyday life of communities. These different goals reflect the different values placed on different kinds of outcome, and these values themselves often draw on further assumptions about human nature or human purposes. Consequently, the arguments in this chapter are to some extent conceptual, concerned with the logical implications or assumptions of particular kinds of 'rehabilitation talk'. The point of this kind of argument is that if our goals are normally taken for granted rather than discussed, they can become confused or incoherent.

Justifications for rehabilitation are essentially moral arguments about what society *ought* to do in relation to offenders, and arguments about what we *ought* to do cannot simply be derived from evidence about what we *can* do: there are plenty of things we can do which we clearly ought not to do. However, there are other kinds of relationship between evidence and justification. One, as moral philosophers frequently remind us, is that 'ought implies can': we cannot reasonably claim that someone has a duty to achieve what is impossible. Another is that different kinds of argument logically require certain kinds of evidence: for example, arguments based on the effectiveness of rehabilitation may require empirical evidence of the changes it produces in offending behaviour, whilst arguments based on the rights of offenders may require demonstrations of consistency with generally accepted principles concerning human rights. Such demonstrations belong to a logically different category of evidence, to which no amount of reconviction-counting could be relevant. This chapter, in identifying different kinds of justification advanced for rehabilitation, is also concerned with the kinds of evidence or argument logically required by each. Later chapters will explore in more detail the empirical evidence actually available to inform some current policy choices: this chapter is concerned with the preparatory ground-clearing task of identifying where different kinds of evidence might, in principle, be relevant. We may have no duty to achieve the impossible, but experience suggests we can waste a good deal of time and money trying.

Prison, drink and saving souls

Histories of the probation service in England and Wales usually start from the Church of England Temperance Society's decision in 1876 to establish a missionary service in certain police courts (see, for example, McWilliams, 1983; Vanstone, 2004). This was an extension of their normal work of trying to persuade sinners, and particularly drunkards, to reform. Ultimately this was for the good of their souls, as well as to reduce the harm they would otherwise continue to do to themselves and others, such as their families. The missionaries' activity clearly belongs in the rehabilitative tradition: a successful outcome was a respectable, self-supporting, abstinent citizen making his way in the world, or a dutiful, thrifty, abstinent wife and mother. The ultimate goal and justification, however, was their spiritual welfare: the successfully helped offender was 'saved' rather than 'lost'. Christians had a duty to show mercy to sinners, and charity gave this a practical form, but active and caring human contact was necessary to persuade sinners and unfortunates to reform. This kind of work was seen as the business of voluntary organisations and charities rather than government. Governments saw their role as safeguarding the conditions for economic development and wealth creation: this they did partly by operating a harsh Poor Law designed to deter idleness and dependency, and a punitive criminal justice system. Meanwhile they generally welcomed the contribution of private and what would now be called faith-based charities, provided that these were not too disruptive or radical.

A good example of how the social and the spiritual were intertwined in missionary work among offenders is provided in the writings of 'General' William Booth, leader of the Salvation Army, who brought together his thoughts on current social problems and proposed solutions in his book *In Darkest England and the Way Out* published in 1890. One chapter of this book, entitled 'The Criminals', is actually about what we would nowadays call the resettlement of prisoners, and the language reveals much about the assumptions underlying such work at the time. The Salvationists had a particular sympathy with prisoners as some of them had themselves served prison sentences, and they were perhaps less 'safe' and more given to social criticism than the Church of England, but the sentiments expressed can be taken as representative of Christian rehabilitative missionary work at the time. Criminals, according to Booth, are sometimes 'hereditary', but their numbers are constantly increased by others who slip into criminality through sheer misfortune: 'Absolute despair drives many a man into the ranks of the criminal class, who would never have fallen into the category of criminal convicts if adequate provision had been made for the rescue of those drifting to doom' (Booth, 1890, p. 58).

Such people may lack the strength or good fortune to reform, particularly when the social reaction to their criminal status denies them the opportunity: 'When once he has fallen, circumstances seem to combine to keep him there…the unfortunate who bears the prison brand is hunted from pillar to post, until he despairs of ever regaining his position, and oscillates between one prison and another for the rest of his days' (ibid.). However, given appropriate help, such a man might 'regain his position' and more. Booth quotes in detail an account written by one recidivist who despaired of finding the means of survival outside prison until: 'In this dire extremity the writer found his way to one of our Shelters, and there found God and friends and hope, and once more got his feet on the ladder which leads upward from the black gulf of starvation to competence and character, and usefulness and Heaven' (Booth, 1890, p. 61). Social and spiritual rehabilitation are presented here as one process, but the overarching goal is salvation, and other achievements are valued mainly as means towards this end. In addition, the pursuit of salvation and the exercise of mercy are matters for individual choice rather than public policy, and mercy assumes that severity is the norm, since mercy consists in not exacting the usual rigorous penalty. Evidence of achievement of the ultimate goal of salvation is beyond the reach of secular social science; on the other hand, 'competence, character and usefulness' are themes to which we will return repeatedly, in various forms, throughout this discussion.

Utility, the State and social reconstruction

As Garland (1985) points out, the early part of the twentieth century was already seeing the emergence of a 'penal-welfare complex' which, among other developments, began to involve the State as a key actor in the business of rehabilitating offenders. No longer was the offender to be rehabilitated to save a soul for God; instead he or she was to be helped towards 'competence, character and usefulness' in the service of the proper collective goals of a secular State – a good citizen rather than merely a good person. This was to emerge most clearly around the middle of the century, when two major wars separated by an economic crisis had led to the development, particularly in Europe and the United States, of forms of government which exercised an unprecedented degree of control over the economic and social life of citizens. They had learned to work together in the common (national) interest, and increasingly expected governments to develop collective solutions to social problems. The dominance of the machinery of government, and the dominant economic role of government expenditure which had developed during the war years, were turned in the 1940s to the new task of social reconstruction through the development of Welfare States (Sullivan, 1996), and the construction of the citizens of the future through publicly funded education. The criminal justice system was only a small part of the system of government but was also touched by this reformist vision, in which new models and methods for the rehabilitation of offenders were enthusiastically advocated and practised. Their language and concepts echo those developed by the American Mental Hygiene movement and the emerging social work profession (Wootton, 1959): offenders were maladjusted, and the point of rehabilitation was to make them normal (good citizens) by appropriate adjustment.

A good example of these arguments is provided by the work of Herman Mannheim, a refugee from Nazi Germany who brought his experience of German jurisprudence and continental criminology with him to Britain and in turn became one of the pioneers of British criminology (Hood, 2004). In one of his books, *Criminal Justice and Social Reconstruction* published in 1946, he sets out a programme for the development of the criminal justice and penal systems in the 'reconstructed' post-war societies. Along with a chapter on making the administration of criminal justice 'more democratic', he provides a set of recommendations for making it 'more scientific'. The aim is a penal system which has the best effects for society as a whole, resembling the traditional utilitarian emphasis on the 'greatest happiness shared among the greater number' (Beccaria, 1963). This is made particularly explicit in Mannheim's discussion of the need for reform of laws prohibiting homosexuality: 'The New Order cannot afford to squander its energy and resources on the persecution of otherwise decent and useful members of the community, merely because their sexual behaviour differs from that of the majority' (Mannheim, 1946, p. 62). The transformation of offenders into decent and useful members of the community by the most efficient means, whether that involves reducing the reach of the criminal law or changing the behaviour of offenders, is a project in the best utilitarian tradition, which always attempted to apply clear principles to the practical business of social administration (Bentham, 1823). It also offers a clear justification for rehabilitative efforts: they are undertaken in the interests of society as a whole, to maximise the availability of 'decent and useful members of the community' for the collective task of social reconstruction.

The means to be used in delivering the new 'scientific' criminal justice were, primarily, new 'expert' inputs into the sentencing and management of offenders. 'Experts' (psychiatrists and psychologists) should advise the court before sentence, and a 'Treatment Tribunal' should be set up, based largely on the model of the California Youth Correction Authority established in 1941, to classify, allocate and if possible rehabilitate offenders placed into its care under indeterminate sentences passed by the courts. Mannheim clearly perceived some problems in relation to traditional ideas of fairness: 'It is no use denying that, in its practical consequences, individualization of treatment, that dominating principle of modern penology, is bound to clash with the traditional requirements of justice as understood by the man in the street' (Mannheim, 1946, p. 228). However, the claims of modernity were not to be denied by such old-fashioned prejudices:

> a partial solution to the problem lies in the working out of really scientific principles of individualization which will make it possible at least roughly to re-establish the rule of equal treatment of equals. It is in this connection, too, that the Prediction Tables worked out by the Gluecks will be valuable. As soon as these new principles become known and accepted, beyond a small circle of experts, by the community at large, individualization will no longer be suspected as injustice. (pp. 228–29)

(The reference to the work of Sheldon and Eleanor Glueck is presumably to their early work, particularly Glueck and Glueck 1934 and 1940, and represents a precursor of recent attempts to use actuarial risk assessment to support rehabilitation – an issue to which we return in later chapters.) As first step in reform, Mannheim proposed the setting up of a central Board to advise the courts and manage the whole institutional side of the penal system; an extension of the scope of indeterminate sentencing; and the prohibition of short prison sentences, with 'consequent strengthening of the probation service' (p. 237).

Variations in the utilitarian approach

The utilitarian justification of rehabilitation as being in the interests of society as a whole has taken a number of forms in its long and influential life, but one of the most obvious and important changes is a periodic shift between what might be called a 'strong and weak' version of the argument, or perhaps more accurately an optimistic and a guarded claim. Briefly, the strong or optimistic claim is that society as a whole benefits from dealing with offenders in such a way as to reduce their offending: rehabilitating offenders contributes to the general good. The weak or guarded claim is that although we cannot be confident in our ability to change offenders for the better, we can at least avoid unnecessary harm resulting from excessive or damaging penalties. This argument is often used, for example, to argue for a presumption in favour of community penalties and against custodial penalties, and often combined with the argument that even if the effects on offenders are similar, the community option is cheaper and so the principle of maximising general benefit applies.

Sometimes this is broadened (for example by Rutherford, 1986, or in a more extreme form by Schur, 1973) into a general argument in favour of lower rather than higher degrees of intervention in the lives of offenders. In essence, however, both arguments are of the same kind: a rehabilitative approach is supported, in the utilitarian tradition, on the grounds that it makes the greatest contribution to the general good of society. The choice between strong and weak forms of the argument depends largely on the state of current opinion regarding the effect of rehabilitative penalties on offenders' behaviour: the strong form of the argument is deployed in periods of optimism about this (Mannheim's proposals are a good example), and the weak form tends to be used in times when people are less confident about the effectiveness of rehabilitative penalties. If nothing works, cheaper is better. In later chapters we review some dramatic shifts which have occurred between strong and weak versions of the case for rehabilitation.

For now it is sufficient to note that that the underlying optimistic or sceptical stances are sometimes influenced by evidence; but not only by evidence; the way in which evidence is received, understood, disseminated and used is in turn influenced by culture, ideology, social change and perceptions of political advantage. In Mannheim's time the evidence that rehabilitative sentences could change offenders was not strong, but optimistic beliefs were supported by the general commitment to improvements in social welfare: as his contemporary Radzinowicz wrote, '[Probation] received a powerful stimulus from the contemporary change of attitude towards the purposes and effects of judicial punishment and from the ameliorative social creed of the Welfare State' (Radzinowicz, 1958, p. xi). In recent times the empirical evidence has been stronger, but the political and social climate often less favourable.

In so far as these arguments depend on a utilitarian justification, they are vulnerable to a number of traditional criticisms of utilitarian theories of justice. These have recently been reviewed by Hudson (2003), and several of them are relevant to arguments about rehabilitation: for example, if rehabilitative penalties are held to be generally beneficial to society, does it matter if the offender who is being rehabilitated is not actually guilty of the offence, provided that he or she is generally believed to be guilty? One version of utilitarian theory is often believed to imply that punishment of the innocent is justified if it contributes to the general good, which clearly conflicts with widely accepted views about justice. The usual defence advanced against this argument by those sympathetic to Utilitarianism (see Urmson, 1953; Rawls, 1972) is that arguments about general benefit are concerned with social practices or rules governing action rather than with individual actions: from the perspective of this so-called 'Rule Utilitarianism' it is in the general interest to adhere as far as possible to a rule that the innocent should not be punished, since such a rule is necessary to maintain confidence in criminal justice and protection from arbitrary punishment.

Other criticisms are perhaps more telling: for example, if we are applying a 'greatest happiness' principle, might not executing some offenders make more people happy than rehabilitating them? If revenge is satisfying, why not provide it? These examples point to a need to distinguish between a 'greatest happiness' principle and what might be called a 'greatest benefit' principle, or between short-term wants and long-term interests, which suggests that the kind of justifications of action required to make sense of utilitarian arguments are both more complex and more debatable than its original advocates thought. This is also a consequence of other long-standing criticisms of utilitarian justice which raise questions about individual rights, the instrumental use of human beings and what offenders deserve.

Rights, needs and 'treatment'

Questions about individual rights are most clearly raised by the example of conviction of the innocent mentioned above, and they also inform arguments in favour of many aspects of procedural justice and due process such as the right to a hearing, to know the charges, to present and contest evidence, to be presumed innocent until proven guilty, and not to be tortured. The contemporary importance of these arguments is shown by the willingness of governments to ignore national and international laws designed to protect such rights when they are seen as inconvenient obstacles to some higher purpose such as the 'war on terror'. In the field of rehabilitation, arguments about rights are more likely to emerge in the less dramatic contexts of debates about consent to the imposition of court orders (originally, and still in many countries, a standard feature of probation orders, but abolished in England and Wales because it was said to detract from the dignity of the court). This is a particular case of a wider concern about offenders' right not to be subjected to arbitrary and indeterminate 'treatments' simply because somebody has decided this will do them good.

The topic of 'treatments' and 'treatment models' and the question of whether action taken to rehabilitate offenders can properly be understood as 'treatment' will recur throughout this book, but some aspects of it require discussion here as they are particularly relevant to questions of justification. Treating people who are ill is normally seen as a proper and desirable activity, and advocates of rehabilitation have often tried to draw on these arguments both to add legitimacy to their efforts and to assert that the interests and needs of individuals lie at the heart of their approach. Again Radzinowicz provides a clear example: '...probation is fundamentally a form of social service preventing further crime by a readjustment of the culprit...' (Radzinowicz, 1958, pp. xi–xii).

This had also emerged as a professional consensus: for example, the European Seminar on Probation held in London in 1952 heard a paper from an official of the United Nations Secretariat on 'Probation and its place in a rational and humane programme for the treatment of offenders' (Pansegrouw, 1952). This informed the delegates that 'probation is individualized treatment, and effective probation practice presupposes the intelligent use of scientific knowledge and techniques both in selection for treatment and in the treatment itself... probation is supervision or treatment in the community', although the lecturer recognised that 'at present, probation supervision and treatment in different jurisdictions still ranges from religious counselling or the unskilled advice of volunteer probation officers, on the one hand, to professional social case work, sometimes supplemented by psychiatric treatment, on the other' (Pansegrouw, 1952, pp. 12–13). Here we see, at quite an early date, two of the themes which were to be central to the growth of social work during the 1950s and 1960s: social workers are professionals working scientifically, and their intervention is justified by the needs and interests of individuals. In this way they began to move the emphasis of their work away from utilitarian concern with the implementation of social policy towards a claim that individual needs and relationships were their primary focus.

This assertion of the central importance of individual needs was intended partly to guard against the utilitarian tendency (exemplified particularly by the authoritarian collectivist regimes then dominating Eastern Europe) to subordinate individual interests to collective requirements. Early classics of social work writing (such as Biestek, 1961) vigorously defended the individual focus. However, this stance in the emerging social work profession also attracted strong criticism for its neglect of a sociological understanding of social problems (Mills, 1943), its indifference to collective social policy initiatives (Sinfield, 1969) and its lack of an explicit agenda for radical social reform (Bailey and Brake, 1975). In the specific context of social work with offenders, criticism tended to focus on the idea of 'treatment': treatment implied illness, which was different from crime and typically involved involuntary incapacities rather than chosen misbehaviour (Flew, 1973); treatment typically involved doing things to a passive and objectified patient, rather than collaborating with an active and reasoning person to solve problems (Bottoms and McWilliams, 1979), and treatment attributed to individual pathology problems which were actually consequences of social disadvantage (Walker and Beaumont, 1981). Moreover treatment was, in other fields and professions, voluntary except for those demonstrably unable to decide for themselves: this was particularly true in those professions which social work aspired to emulate, such as medicine and psychotherapy.

Interesting mental gymnastics were performed in the attempt to assimilate supervision of offenders under a court order to the model of a voluntary therapeutic relationship. For example, Foren and Bailey (1968) argued that a degree of coercion of the 'immature' was justified by the fact that, when helped to become 'mature', they would appreciate how necessary it had been: a letter from a Borstal trainee to his probation officer thanking him for recommending a custodial sentence (Hunt, 1964) was quoted in support, and this no doubt rather untypical event had to carry the burden of justifying a curiously hypothetical and retrospective reframing of the concept of consent. Others argued that effective supervision of offenders was essentially a negotiated process in which offenders participated and made choices: 'help' rather than 'treatment' (Bryant *et al.*, 1978; Raynor, 1985).

Rehabilitation as a right

These arguments about the nature of 'treatment' and the status of its human subjects resonated with other criticisms of utilitarian aims in sentencing which raised concerns about the instrumental use of people. Just as 'treatment' tended to reduce people to passive objects of intervention, so an approach to sentencing which chose penalties primarily on the basis of their general social effect seemed to involve an instrumental use of people as means to goals which were not their own. This seemed to conflict with Kant's argument (1965) that because people are moral beings capable of choice, they must not be treated simply as instruments of other people's purposes. Sentencing directed purely to social goals seemed also to neglect individual human agency in other ways: recognising people as moral agents implied recognising desert and censure as elements in sentencing. This argument was advanced both in a religious moralistic form (people had a right to punishment which recognised their moral responsibility and offered opportunities for atonement – see Lewis, 1971) and in a secular liberal form, which argued that justice in sentencing involved looking backwards to assess what the offence deserved (essentially a moral assessment) rather than looking forward to the possible prevention of future offences (American Friends Service Committee, 1971; von Hirsch, 1976). A typical statement of this position, which intended to limit punishment by contesting the excesses of disproportionate 'treatment' and preventive or deterrent incarceration, was provided by Roger Hood:

> I believe a system which arrives at the length of sentences based more on a moral evaluation than on appeals to the utilitarian philosophy of deterrence and reductivism, would be fairer, not necessarily less effective, possibly less, not more punitive and appeal to that sense of social justice on which any acceptable system of social control must be founded. (Hood, 1974: 7)

Contemporary penal policies advanced under the banner of 'just deserts' fall short of these liberal intentions in many ways (Hudson, 1987): however, from the point of view of justifying rehabilitation, such arguments suggest some additional tests which rehabilitative efforts would need to satisfy. Briefly, if they are part of a penalty they need to be proportionate, or at least should not constitute a greater degree of intervention than the seriousness of the offence merits. If not part of the penalty, they should be voluntary (like post-release aftercare for prisoners released without a compulsory licence). They should not inflict damage or detriment beyond that specified as part of the sentence (a real issue given the evidence that some intended rehabilitative efforts actually have negative effects – see, for example, McCord, 1978; Walker *et al.*, 1981). Most fundamentally, they should be compatible with an understanding of offenders as morally responsible for their actions unless demonstrably incapable of choice through *force majeure* or independently evidenced incapacity, such as severe mental illness.

Thus the introduction of arguments about rights and moral responsibility allows a powerful critique of models of rehabilitation which relied on one-sided treatment and utilitarian justifications. However, they also open the door to some other kinds of justification. For example, if justice requires that penalties should be determined by seriousness and culpability, should not offenders whose circumstances offer them few alternatives to crime be treated as less blameworthy than those who have many resources, options and alternatives? Consider the impoverished single mother who makes a false statement to support a benefit claim when her children do not have enough to eat. Is she as blameworthy as the wealthy businessman who makes a false statement to secure some financial advantage simply to satisfy greed? It has been suggested that in order to recognise these differences of circumstances, opportunities and power, just deserts approaches to sentencing should allow a 'hardship defence' which partly or wholly mitigates the penalty (Hudson, 1999). This would also be relevant to people who offend because they are threatened or coerced by others who then benefit from the proceeds. What is relevant to the current discussion is that a recognition of hardship and of unequal opportunities to avoid crime suggests not simply mitigation of the penalty, but also that a State which seeks to guarantee a minimum acceptable standard of living and level of welfare to its citizens is obliged to offer to offenders the services which could make avoidance of crime a more realistic prospect.

This approach to justifying rehabilitation has become known as 'state-obligated' rehabilitation (Cullen and Gilbert, 1982; Rotman, 1990; Carlen, 1994), and rests on a version of social contract theory: the moral legitimacy of the State's demand that people refrain from offending is maintained if the State fulfils its duty to ensure that people's basic needs are met. Welfare States are the most familiar modern version of this social contract, and rehabilitation is not simply justified but mandated by the clear connection between social deprivation (or, in more modern jargon, social exclusion) and offending. Rotman also argues that there is a duty to provide rehabilitation to mitigate damage done by punishments such as imprisonment. Thus the obligation to meet needs is justified not simply by the Kantian appeal to the importance of people as ends in themselves (which McWilliams [1987] described as the 'personalist' approach to rehabilitation) but by a political theory of the duties of States and citizens to each other. As Rotman puts it,

> rehabilitation becomes a right of offenders to certain minimum services from the correctional services. The purpose of such a right is to offer each offender an opportunity to reintegrate into society as a useful human being. (1990, p. 6)

Later chapters will explore some of the implications of this in more depth.

Rehabilitation for the benefit of potential victims

Other current models of rehabilitation, particularly those based on social learning theory and often delivered through 'programmes' (McGuire, 1995, 2002; Raynor and Vanstone, 1997), aim to empower offenders to take more control of their lives and behaviour and to make more pro-social choices by helping them to learn necessary skills such as listening and communication, critical and creative thinking, problem-solving, self-management and self-control. Such approaches recognise problems in relation to resources and opportunities but see little point in improving access to these without also ensuring that people have or develop the necessary skills to benefit from them.

In general these approaches to rehabilitation have taken the issue of evidence very seriously, unlike earlier approaches, but they have sometimes been less clear about their philosophical and theoretical base. Are they 'treatment' or social learning? Do they remedy deficits, or enhance freedom and choice? Such issues will be discussed further in other chapters, but for now it is important to note that the pioneers of these approaches have often been psychologists, and impatient of metaphysical or theoretical speculation which they see as going beyond the evidence (Andrews and Bonta, 1998). As a result their methods have been criticized (in our view wrongly) as a revival of the discredited 'medical model' of treatment (Mair, 2004); but what is, in technical and practical terms, a fairly new and very promising approach to rehabilitative work has been content to rely on traditional justifications. Rehabilitation is advocated (for example by McGuire, 1995) on the grounds that it is better for both offenders and society because it can reduce further offending and victimizations. Here we see again a utilitarian appeal to the general good, and it is noticeable that work of this kind with offenders in practice usually gives priority to public safety through the use of risk assessments. This primacy accorded to public safety is described by Garland (2001) as a shift in the justification of rehabilitation: the emphasis, he argues, has moved from the benefit to the offender towards the benefit to potential future victims – it is for their sake that rehabilitation is attempted.

Rehabilitation for the benefit of communities

Finally, some recent approaches to justifying rehabilitation have begun to make use of a concept of community which, instead of excluding the offender, includes him or her as part of the community of interests to be addressed. Most of the arguments we have reviewed up to this point set the offender and 'community' or 'society' against each other, as if the offender is not part of a community which consists only of his or her potential victims. Hence the offender's interests are always counter-posed to those of the 'community' and weighed against them, or assumed to be in conflict. Of course there are areas of conflict: however, just as state-obligated rehabilitation is based on the rights that offenders share with other citizens even after they have offended, communitarian approaches to rehabilitation recognise that offenders mostly belong to communities, and that their memberships and affiliations need to continue, or to be repaired, if they are to be reintegrated into normal membership of communities. Such approaches are associated particularly with advocates of restorative justice (for example Braithwaite, 1989) who believe that reintegrative processes can help offenders to atone for or make reparation for their offences at the same time as helping offenders and victims to acknowledge the wrong and to learn something of each other. The aim is the restoration or establishment of social bonds that will both offer the offender membership of a community and consequently strengthen informal controls over his or her behaviour. Whilst some of these ideas properly belong in a discussion of restorative justice rather than rehabilitation, the fact that offenders are meant to learn from restorative procedures a social lesson which will influence their future behaviour places them also within the territory of rehabilitation.

One implication of this is that rehabilitation should be seen not simply as meeting offenders' needs or correcting their deficits, but as harnessing and developing their strengths and assets. The 'Justice Reinvestment' movement in the United States sees poor communities as doubly deprived by the imprisonment of many of their young and energetic members, who should be active as producers, earners and parents within the community: it advocates the use of non-custodial penalties with reinvestment in the community of money saved by not imposing prison sentences. A related approach to rehabilitation is also emerging, known as a 'strengths-based' approach (Maruna and LeBel, 2002) which justifies rehabilitation on the basis of the contribution the rehabilitated offender can make to the community, and the community's need for this contribution. 'Strengths-based and restorative approaches ask not what a person's deficits are, but rather what positive contribution the person can make' (Maruna and LeBel, 2003, p. 97).

Summary and conclusion

In this chapter we have considered a number of justifications historically advanced for a rehabilitative approach to offenders. These justifications are not primarily concerned with evidence (what *can* be done) but with the obligations and duties of individuals, societies or communities (what *ought* to be done). Each reflects the assumptions and current knowledge of its time, but versions of each can still be found in contemporary discourse, and their differences and contradictions can lead to confusion. Each also logically requires a different type of evidence to support it, and in most cases the weight of the evidence is less than an enthusiastic advocate would wish. Other chapters in this book will be centrally concerned with assessing that evidence.

We have seen how an early faith-based commitment to saving souls and the exercise of mercy, which originally had little to do with public policy, was changed by the emergence of interventionist social policy and Welfare States. The scientific 'treatment' of offenders was justified not only by reference to the common good, but also by the emerging social work profession's commitment to individual needs as understood within its own diagnostic framework – rehabilitation as 'treatment'. A renewed emphasis on the rights of offenders as moral agents, and the moral requirement to deal with them as their past behaviour deserved rather than as their needs or their expected future behaviour required, began by undermining the vision of rehabilitation as one-sided professional intervention. In due course, however, the focus on rights gave rise to a new conception of rehabilitation as a service the State is obliged to provide to offenders who have been unfairly disadvantaged, and therefore have had restricted opportunities to avoid crime.

Instead of 'treatment', a 'learning' model of rehabilitation emerged, which recognised offenders as moral actors and aimed to help them to acquire the skills and resources which would help them to make choices which better served their own interests and the interests of others. At the same time a renewed emphasis on risk and public safety meant that rehabilitation was increasingly seen as justified not so much by its beneficial effect on the lives of offenders as by improvements in the safety of potential future victims. A new 'treatment' language (which is actually about participative social learning rather than one-sided 'treatment' of passive subjects) aims to justify rehabilitation by reference to demonstrable changes in offenders' behaviour, but still leans towards an emphasis on 'deficits' and 'correction'. Meanwhile other supporters of rehabilitation seek to resolve the historic tension between the interests of offenders and 'society' by pointing to their relational connection in communities. Offenders need communities, and communities need rehabilitated offenders: rehabilitation is enjoined on society not simply by offenders' needs or deficits, but by their strengths, assets and potential contribution (an argument which would probably not have surprised William Booth).

What is clear is that arguments in support of rehabilitation necessarily involve continuing debates about the balance between rights and needs, strengths and deficits, desert and risk, individual interests and public policy. Into this debate we add the familiar argument that 'ought implies can': our social obligations to rehabilitate offenders depend on the question of how far it is actually possible to do so, and this is inescapably a matter of empirical evidence.

3
Origins and Contexts

The entry of 'reform' into penal discourse and practice in Europe has been the subject of a number of authoritative historical analyses. In these accounts, the period between the late eighteenth and the mid-nineteenth centuries is identified as a key turning point in thinking about how best to deal with offenders. In the work of both Foucault (1977) and Ignatieff (1978), the period between 1775 and 1850 is highlighted as inaugurating the reform or 'correction' of offenders as a legitimate and practicable penal objective. For both writers, the evolution of new ideas about the reform of offenders is understood as part of a qualitative shift in systems of penality which occurred in this period, a shift which is characterised by the displacement of deterrent systems of punishment based on ceremonial displays of sovereign power, in favour of systems of carceral punishment operating according to predictable rules and the tenets of the human sciences as they were understood at the time. In Foucault's *Discipline and Punish* this is graphically represented by the displacement of the scaffold by the prison; in Ignatieff's work the construction of Pentonville prison takes centre stage.

It is clear that in the limited space afforded in this chapter we cannot do justice to the 'pre-history' of offender rehabilitation: all we can hope to accomplish is an outline of some of the key ideas about correction and reintegration which were contemporary in the eighteenth and nineteenth centuries, and consider the contexts from which they emerged. In the first part of this chapter we focus on the emergence of ideas about how the prison might be made to work in the interests of offender rehabilitation. We then move on to consider parallel ideas about rehabilitating offenders beyond the walls of the prison or penitentiary and the 'formal' penal apparatus, namely in the community context. We conclude the chapter by considering the publication of the report of the

Departmental Committee on Prisons (1895), commonly known as the 'Gladstone Committee Report', which is widely regarded as a pivotal moment in the history of the 'reformative ideal'.

Visions of 'penitentiary discipline'

Prior to the sixteenth century, the monastery was the only context in which imprisonment, or 'penitential confinement', had been deployed with a view to achieving correctional or 'reformatory' effects (Pugh, 1970; McConville, 1981). Pugh explains that, whilst it is not clear precisely when monastic prisons grew up, there is documentary evidence to suggest that by the ninth century they were commonplace in parts of Europe, and in England by the fourteenth century. Historians also know little about the kinds of offences which warranted such confinement, although Pugh cites examples of theft, forgery and violence; and, in the case of one Yorkshire nun, bearing a child (Pugh, 1970, p. 381). As McConville has explained, various monastic orders provided for separate confinement for such 'grave offenders' within the cloistered community:

> Monastic prisons isolated the malefactor, partly as duress and punishment, partly in order to reduce moral contagion, but also with the intention of curing the offender's physical and spiritual defects. Such prisoners were kept in silence, subjected to a special diet, and allowed only the distraction of approved books and conversation with their abbot or some designated elder brother. (1981, p. 31)

In Britain, the first secular attempt to undertake 'moral correction' through subjection to an institutional regime was instituted in the mid-sixteenth century. In December 1556, Henry VIII's palace of Bridewell opened its doors to some of London's 'undeserving poor': the class of idle 'vagabonds' whose ranks were thought to pose a demoralising example and to represent a breeding ground for crime. Bridewell was the first of a number of 'houses of correction' to be established in England and parts of Europe in the next few decades. In contrast to the practice of monastic confinement, the regime of the houses of correction centred not on isolation and what was in the context of sixteenth-century English society perceived as 'idle reflection', but rather on the curative potential of productive activity. These institutions were conceived as places where members of this particular worrying class could, through enforced labour, be rendered both productive and 'morally improved'. By 1579 twenty-five different trades were practised at Bridewell, and by 1631 there were sixteen tradesmen teaching their skills to a total of 106 apprentices (McConville, 1981).

Some two centuries later, aspects of both the monastic and the Bridewell regimes were to be revived in the writings of John Howard, the Bedfordshire magistrate and penal reformer who has been characterised not only as the leading figurehead of the 'evangelical' approach towards offender reform (Forsythe, 1991) but also as the father of the 'reformative regimentation of criminals' (Ignatieff, 1978, p. 50). As is well documented, Howard's commitment to penal reform was a spiritual calling discovered when, in his capacity as a Bedfordshire magistrate, he took it upon himself to inspect the local gaol. There he uncovered an unregulated 'netherworld' of ill-health, abuse and extortion which prompted him to undertake a detailed survey of every penal establishment and house of correction in England. First published in 1777, Howard's *The State of the Prisons* culminated in a draft Bill setting out his vision for the reform of both prisons and offenders. At the heart of that vision was the notion of 'penitentiary discipline' (Ignatieff, 1978), an orderly institutional regime designed to bring about the 'moral correction' of prisoners. Howard's prison 'census' also prompted overdue attention to the nutrition and physical health of prisoners.

In considering Howard's vision of penitentiary discipline it is important to acknowledge the particular context from which it emerged. Characterised by Ignatieff as a 'theatre of deterrence', eighteenth-century penality relied heavily upon the public rituals of hanging, whipping and pillory, increasingly supplemented as the century progressed by transportation to America and the developing colonies (Rutherford, 1998). Although sentences of imprisonment could be passed by magistrates in relation to summary offences, prison was almost exclusively deployed as a place of confinement for debtors[1] and those awaiting sentencing or punishment – most commonly, transportation or execution (McConville, 1995a). When used as a punishment, imprisonment was solely retributive and deterrent: prison, in common with the other eighteenth-century punishments, had no discernible reformative aspect (McConville, 1981). But by the mid-eighteenth century, doubts were increasingly being voiced about both the legitimacy and the deterrent effects of the existing range of punishments. Among those, a generation before Howard, to voice serious concerns about the overly harsh and otherwise inadequate range of existing penalties available to mid-eighteenth-century sentencers was the magistrate, novelist and pamphleteer Henry Fielding. In one of two ambitious 'social pamphlets' in which he addressed the problem of the 'idle poor', Fielding (1751) argued that it was necessary to find an alternative to hanging for petty offenders, the objective of which should be to render them 'useful members of society'.

It is clear from Howard's writings that Fielding's ideas were one source of inspiration for his own model of penitentiary discipline. Certainly Howard shared Fielding's opinion that capital punishment ought to be reserved for only the most serious offenders.[2] Howard's central proposal was that the prison could be made to work in the interests of offender reform. 'The penitentiary houses', Howard remarked, '...may, under proper management, be made to answer very useful purposes' (1929, pp. 44–45). But in order to achieve reform, significant changes in prison conditions were necessary. Again, Howard echoed Fielding's firmly held belief[3] that exposure to religion would go a long way towards correcting the morals of offenders and, to this end, he proposed a central role for prison chaplains, envisaging both regular sermons and bible readings for all prisoners. In the spirit of the monastic disciplinary practices described above, Howard was also a proponent of the restorative power of reflection. But for Howard, solitude and silence were not only the precursors of repentance; they were also antithetical to the problem of 'moral contamination' which he observed in the course of conducting his prison census, and it was for these twin reasons that these ingredients were viewed as essential to a reformative regime.

Another particular source of inspiration for Howard was his visits to the houses of correction ('Rasp Houses') of Amsterdam and Rotterdam. Two aspects of the practices he observed in these institutions particularly impressed Howard. First, already convinced that men could be changed by 'awakening their consciousness of sin' (Ignatieff, 1978, p. 67), Howard was clearly impressed by the centrality of religious instruction in the Rasp Houses. As he observed,

> Great care is taken to give them moral and religious instruction, and reform their manners...The chaplain (such there is in every house of correction) does not only perform public worship, but privately instructs the prisoners, catechises them every week, etc., and I am well informed that many come out sober and honest. (Howard, 1929, p. 48)

Following his account of Dutch practices Howard makes clear his 'ardent wishes',

> that our prisons also, instead of echoing with profaneness and blasphemy, might hereafter resound with the offices of religious worship; and prove, like these, the happy means of awakening many to a sense of their duty to God and man. (1929, p. 53)

A firm believer in the immorality of idleness, Howard was also convinced of the reformative effect of enforced labour, and approved of the European practice of putting convicts to work as an alternative to transportation. For Howard, the replacement of idleness by the 'habits of industry' was essential not only to the moral improvement of offenders, but also to their future social utility. Work, then, he viewed as necessary 'in order to correct the faults of prisoners, and make them for the future more useful to society' (1929, p. 40).

Although inspired largely by his Christian beliefs, Howard's ideas fell on particularly fertile ground, coinciding with the ascendancy of the Utilitarian penal philosophy of Cesare Beccaria, whose essay *On Crimes and Punishment* first appeared in English in 1764. Although not directly concerned with the reform of offenders, Beccaria's assertion that the harms associated with punishments need only marginally exceed the perceived benefit derived from the crime came not only to constitute a powerful critique of eighteenth-century penality, but also indirectly lent support to calls, such as Howard's, for more humane and effective criminal sanctions (McConville, 1981).[4] Ignatieff explains how both Howard's belief in the possibility of reforming offenders and his 'disciplinary solution' to the problem of crime also dovetailed rather neatly with materialist ideas which became influential during the 1770s. Indeed, Ignatieff suggests that the rise of materialist beliefs provided the context necessary for the acceptance of Howard's ideas about prison discipline. Derived largely from the work of John Locke and David Hartley, the 'scientific' tradition of English materialism denied the existence of innate propensities or ideas and theorised criminality as a result of incorrect socialisation. In this way, materialism offered a 'scientific' rebuttal of the notion that offenders were incorrigible and proposed in its place a vision of offenders as 'merely defective creatures whose infantile desires drove them to ignore the long-term cost of short-term gratifications' (Ignatieff, 1978, p. 67). By collapsing the distinction between the body and the mind, materialist psychology also seemed to offer a scientific explanation for Howard's claim that men's 'moral behaviour' could be altered by disciplining their bodies. Materialist psychology theorised that a disciplinary regime applied to the body would initially become a habit and then gradually be assimilated, ultimately becoming an automatic part of the person's behavioural repertoire.

This approach to reform was exemplified in Bentham's infamous 'Panopticon' design (Bentham, 1791). Modelled on a factory which his brother had built in Russia, Bentham's Panopticon entailed a circular building with cells around its perimeter and a central inspection tower.

The architectural design ensured that, whilst each cell and its occupant would, by virtue of windows and back-lighting, be clearly visible from the central point, the interior of the tower itself would remain opaque to the cells' occupants. Thus the presence or absence of the supervisor or inspector would be unverifiable, but the effect would be that of permanent visibility and constant surveillance. In this way, Bentham believed, the technology of the prison would render physical constraints unnecessary.

Howard's notion of penitentiary discipline, then, held obvious attractions for the more secular reformers. However, it is important to note, as Ignatieff does, that this common belief in the possibility of reform and the concurrence as to the means or technology for bringing about reform in fact masked two very different 'criminologies':

> Howard and Bentham both denied criminal incorrigibility, but from diametrically opposed positions – one accepting the idea of original sin, the other denying it. One insisted on the universality of guilt, the other on the universality of reason. Materialists like Bentham...asserted that men could be improved by correctly socializing their instincts for pleasure. Howard believed men could be changed by awakening their consciousness of sin. (Ignatieff, 1978, p. 67)

A shared commitment to the reformative power of penitentiary discipline also masked some subtle but important differences in respect of the *details* of a disciplinary regime envisaged by the two men. Although Bentham's Panopticon design was much influenced by Howard's work, Bentham's vision was essentially that of a capitalist enterprise; a 'prison factory' in which prisoners would work in their cells for up to sixteen hours a day.[5] As such he emphasised productive labour and profit-making over some of the issues closer to Howard's heart, such as the necessity of moral and religious instruction.

Penitentiary practice in the nineteenth century

By the late eighteenth century, then, a growing body of opinion favoured more proportionate, and potentially reformative, penal sanctions, and it was in the context of this growing consensus that a government commitment to the development of a national system of penitentiaries was rendered possible (McConville, 1981). But the 1779 Penitentiary Act,[6] developed in consultation with Howard, was never implemented, and plans to develop two national penitentiaries, rendering long-term imprisonment an alternative to transportation, had been abandoned by 1785 (Ignatieff, 1978; Rutherford, 1998). Ignatieff has suggested that one of the principal reasons why the Penitentiary Act had such a limited impact was ministerial reluctance to accept, as an alternative to transportation, a system of imprisonment which would ultimately return convicts to their own communities. For a society that had long assumed offenders to be incapable of reform, Ignatieff argues, such a commitment would have required a huge leap of faith.

Nonetheless, the Act and the ideas behind it did prompt a spate of local prison Acts which enabled not only the reform of certain county prisons and houses of correction, but also the construction of a small number of new, local 'penitentiary prisons', such as the one erected in 1791 in Gloucestershire under the management of Sir George Paul. When in 1810 the House of Commons was again persuaded to consider proposals for alternative modes of punishment, Sir George was one of three principal witnesses called by the Holford Committee, along with the reverend John Becher of the Southwell house of correction, and Bentham, who sought to renew interest in his Panopticon. As McConville (1981) has explained, the systems recommended by the three men prioritised and utilised labour, seclusion and religion in different ways, reflecting different theories of criminality and differing views as to the balance of reformative and deterrent objectives. For example, although absolute seclusion had been central to Bentham's original design, he had by now come to modify his views, and suggested that whilst an initial, temporary period of seclusion might be effective in order to 'break the spirit' of certain prisoners (a practice used in the Southwell house of correction), it was otherwise an excessively expensive measure and counter to his vision of a prison factory. Paul, however, was critical not only of Bentham's willingness to forego the reformative effects of solitary confinement in the interests of profit-making, but also of Bentham's neglect of religious instruction. For Paul, prisons were not factories, but places of religious reformation (Ignatieff, 1978, p. 112). Of the three regimes, Gloucester's placed the greatest emphasis on religion: 'solitude, sparse diet and unrewarded labour were all intended to make the prisoner malleable; religion was the reformatory fixative' (McConville, 1981, p. 130).

Ultimately, the Holford Committee was persuaded that many offenders could be reformed by a system of imprisonment which featured seclusion, employment and religious instruction and to this end it approved the construction of a national penitentiary, large enough to accommodate several hundred prisoners.[7] In rejecting Bentham's plan, the committee noted among other concerns that such a scheme would not pay adequate attention to religious instruction and 'moral improvement' and would therefore not constitute a fair trial of a reformative measure. In the spirit of this commitment to offender reform, it was decided that the new penitentiary should receive only those convicts considered capable of reform, namely the young, first offenders and others considered to be as yet not wholly corrupted. In 1816 Millbank, situated on the bank of the Thames, received its first prisoners.

However, Millbank's role as an instrument of reform was to be short-lived. In accounting for its demise as a reformatory institution, Ignatieff stresses the dual problems of recruiting honest and reliable staff and of prisoner rebellion against the strict regime of solitude, hard labour and meagre rations. As Ignatieff explains, the Millbank prisoners' bread riots of 1818 encouraged the myth that the regime was indulgent and unduly deterrent, and there followed a hardening not only of opinion but also of practice on issues of punishment. During the 1820s, although there was still no uniformity in prison discipline, the general trend was in favour of reduced diets and the introduction of hard, punitive labour (such as the treadwheel) and, in the following decade, the use of both solitary confinement and the so-called 'silent system' became influential in English prisons. The culmination of these developments, argues Ignatieff, was the construction of Pentonville in north London, which received its first prisoners in 1842.[8]

But whilst the strict regime of solitude and silence, monotonous labour and religious instruction initially appealed to proponents of both reformation and severity,

> Faith in the reformative promise of Pentonville barely survived the 1840s. In the next decade it became fashionable once again to insist that the 'dangerous classes' were incapable of reformation. (Ignatieff, 1978, pp. 200, 204)

Other historians concur that after 1850, the reformatory objective in penal policy went into decline (e.g. McConville, 1981; Forsythe, 1991). Forsythe draws particular attention to the spread in the latter half of the nineteenth century of ideas about heredity,[9] which undermined optimism about the reform of offenders. In 1863 the Carnarvon Committee, set up to consider the relative merits of deterrence and reform, came down firmly in favour of the former, and indeed argued that regimes intended to bring about reform were inimical to deterrence. By the 1870s, McConville argues, the severity of imprisonment had reached such a level that it could not be increased without endangering the health of prisoners. Indeed, he contends that the severity of prison regimes were such that the new sentence of penal servitude for habitual offenders, which had replaced transportation in the 1850s, was considered by many to be a more lenient sentence than that of imprisonment, despite the fact that it was generally of longer duration.[10] Meanwhile religion, previously at the centre of debates about prison discipline, was in the second half of the century conspicuously absent from discussions of penal policy and practice.

The decline of reform as a primary penal objective was underlined in 1877 by the appointment of Edmund Du Cane as chairman of the Prison Commission – a post which he occupied until 1895.[11] Du Cane's 'doctrine' was, first, that general deterrence should be prioritised over both individual deterrence and reform. If reform could be achieved within a system designed principally to effect deterrence, then all well and good; but in no way was an objective of reform to compromise a regime designed to deter criminal behaviour (McConville, 1981, 1995b). Du Cane's second tenet was uniformity of treatment: the executive was to be afforded no discretion in delivering the sentences meted out by the courts, and prisoners were not to be treated differently on the basis of character or circumstances (Radzinowicz and Hood, 1990, p. 529).[12] Under Du Cane's rule, a man sentenced to penal servitude continued to suffer separate confinement for nine months, and to serve the remainder of his sentence at one of the public works prisons; but his comfort, diet and propensity to earn rewards for good conduct were all reduced. Meanwhile, punishments for breaches of discipline both increased in severity and became more commonplace (see Radzinowicz and Hood, 1990, pp. 531–72 for a full account).

However, Du Cane's 'formula' was to come under attack in the early 1890s when a national newspaper launched a tirade of criticism against both Du Cane himself and his grim regime. A parallel critique in another newspaper, penned by a prison chaplain, took a slightly different critical approach, pointing specifically to high rates of recidivism among ex-prisoners. According to Radzinowicz and Hood (1990), this was the first occasion on which the national media played a major part in destroying the legitimacy of the penal system in the eyes of the public. The outcome of this intense media interest was the appointment, in May 1894, of a Departmental Committee of Inquiry, chaired by Herbert Gladstone, to address the critiques and review the principles and machinery of the penal system. The Committee's conclusions are considered at the end of this chapter.

Beyond the prison walls

It is clear from the foregoing discussion that, in the nineteenth-century context, the majority of debate about reform centred on the prison and the treatment of prisoners, and the focus of activity was the design and implementation of experimental regimes of various kinds. However, in considering the evolution of a rationality of reform in penal policy and practice, the prison was not the only significant site of change. In this section we look beyond the prison walls, to consider the community context.

Early resettlement: The 'ticket of leave'

We have already noted Ignatieff's suggestion that one of the principal reasons why the 1779 Penitentiary Act had such a limited impact was ministerial reluctance to accept, as an alternative to transportation, a system of imprisonment which would ultimately return convicts to their own communities. The issue of 'resettlement' did not in fact come to a head until the 1860s, by which time imprisonment had become the sanction of first resort for the majority of major crimes which had formerly been punished by hanging or transportation (McConville, 1981, p. 419). For the first time, convicts were to be released on home territory, their release administered via a 'ticket of leave' system, an early incarnation of parole (Shute, 2003). This required ex-convicts to report to the police at regular intervals, maintain a steady job and avoid association with other offenders. As Ignatieff explains,

> The first ticket-of-leave men were released at home in 1853 amid general public panic. Almost immediately they found themselves blamed for every crime, large or small. While the director of convict prisons, Joshua Jebb, insisted that the recidivism rate for ticket-of-leave men was lower than those released from county and borough prisons, the ticket-of-leave men bore the brunt of the public's anxiety at the fact that they could no longer count on transportation to rid England of its most serious criminals. The ticket-of-leave men found themselves barred from most employment, harassed by the police, and vilified in the press. (1978, p. 201)

In the context of the ticket-of-leave system, then, there were concerns that, far from encouraging ex-convicts to desist from crime, the new system of licensing was in fact making it more difficult for many to secure and keep honest employment. However, the official line taken was that this was an unavoidable side effect of the system (McConville, 1981, pp. 422–23). Fortunately, by the mid-1860s public anxiety had begun to subside, not least thanks to a reduction in the rate of criminal committals (Ignatieff, 1978, pp. 203–04).

At the same time, the ending of the system of 'social elimination' was not greeted by all as a progressive step. That the cessation of transportation abruptly ended the chance of starting a new life overseas and disappointed at least some offenders is evident in that, between 1853 and 1859, a small group of female convicts in Brixton prison waged a protest against the conversion of their sentences from transportation to imprisonment (Ignatieff, 1978, p. 203). Indeed, McConville notes that as late as the 1870s, 'there remained [...] a certain nostalgic regard for transportation, and in particular for the opportunity which it had offered for "making a fresh start" ' (1981, p. 425). Thus, although clearly not conceived *primarily* as a reformative measure, the potential of transportation to work in the long-term interests of offender reform has been noted by a number of penal historians:

> The possibilities for reformation were not seen as incompatible with the objectives of elimination, retribution or deterrence. By comparing the offender's chances of readjustment in society when released from prison at home with those open to him after the expiration of his sentence in Australia, the balance was definitely in favour of the latter [...] Reformation, fostered by the natural influences of the environment, appeared to be so much more promising and tangible than the lofty, but inevitably remote, hopes of achieving moral regeneration at home. (Radzinowicz and Hood, 1990, pp. 472–73)

However, this is not to imply that the reality of transportation necessarily measured up to the expectations of those who saw it as offering a positive opportunity to start a new life, particularly for women who were commonly sexually exploited both on the convict ships and on arrival in the penal colonies (Zedner, 1995).

Aid for discharged prisoners

Meanwhile, for prisoners discharged on home territory, private (charitable) assistance was an option for some, and magistrates were empowered by Peel's 1823 Gaol Act to make 'moderate' payments to enable destitute prisoners to return to their families or to take up employment (McConville, 1981, p. 362). By providing financially for released prisoners in this way, it was hoped that they would be better able to avoid association with the 'criminal classes'. In the 1850s the provision of such assistance was widened with the formation of discharged prisoners' aid societies. Financed initially by voluntary subscriptions and, from 1862 by grants from public funds, these societies were widely supported and perceived as a useful means of helping to discourage recidivism. However, aid was not doled out indiscriminately. Although inevitable problems of classification arose in practice, care was taken not to offend the fundamental Victorian tenet of social policy which attempted to distinguish between the 'deserving' and the 'undeserving' recipients of aid (Radzinowicz and Hood, 1990, p. 605). Another principle which the Victorian philanthropists strove to uphold was that of 'less eligibility'. Introduced by the 1834 Poor Law Amendment Act, this principle meant that the situation of the able-bodied recipient of welfare should not be made as eligible (i.e. attractive) as that of the poorest independent labourer. But whilst the societies tended to believe that very little money should be given to individuals, they did in many cases seek to help ex-prisoners find satisfactory accommodation and employment, thereby encouraging self-sufficiency (p. 609).

The 1860s saw the extension of support to discharged convicts, in respect of whom the societies played an additionally useful role in offering a limited means of surveillance after release – a role which fitted well with the new licensing requirements put in place by the Penal Servitude Act:

> On release the convicts were interviewed by the society in order to determine their intentions. They were given suitable clothing, lodgings and pocket-money (the last from their prison gratuities) and efforts were made to provide employment for them. A system of regular visitation was established. (McConville, 1981, p. 425)

The nineteenth century also witnessed a number of philanthropic endeavours in respect of the provision of temporary refuges or 'hostels' for released prisoners. One such refuge, for female convicts, was opened in 1856 (McConville, 1981). Another was the 'industrial home' annexed to the Wakefield House of Correction, a lodging-house in which discharged prisoners lived and maintained themselves through productive work (Radzinowicz and Hood, 1990, p. 609).

Juvenile offenders and the reformatory system

In keeping with emerging ideas about 'moral contamination' on the one hand, and 'corrigibility' on the other, the nineteenth century witnessed a succession of attempts to separate juvenile offenders from adults and, if possible, to keep them out of prison altogether (Radzinowicz and Hood, 1990). For example, as early as the 1820s some Warwickshire magistrates were committing young offenders to the care of their employers as an alternative to formal punishment (Bochel, 1976; Raynor and Vanstone, 2002). Matthew Davenport Hill expanded on this idea when he was appointed recorder of Birmingham in 1839, handing young offenders over to suitable guardians and requesting reports from specially appointed 'confidential officers' on their subsequent behaviour.

By the mid-1850s the view that juveniles should not be punished alongside adults and that reform rather than punishment ought to be central to their management had resulted in legal recognition for the reformatory school system. A pivotal influence in the movement to establish reformatory institutions for juveniles was Mary Carpenter, an evangelical philanthropist who campaigned to replace penal incarceration for the young with detention in home-like reformatories, where they might receive appropriate moral education. Although a juvenile penitentiary had been established at Parkhurst in 1838 for young convicts sentenced to transportation,[13] Carpenter was a vocal critic not only of its harsh regime but more generally of the principle of imprisonment for juveniles. Ultimately, however, Carpenter's vision was subject to compromise, on a number of levels. First, the reformatory system did not wholly replace the penal system for juveniles: the Youthful Offenders Act of 1854 empowered courts to send certain offenders under the age of 16 to a reformatory school on completion of a prison sentence of at least 14 days. Second, the impact of the Act was also limited in that magistrates could continue to send juveniles to prison rather than to the new reformatories if they so desired, and young offenders frequently spent long periods in prison before reformatory places could be found (McConville, 1981). Third, although some schoolwork and religious instruction were common features of reformatory regimes, hard labour was a consistent and substantial ingredient of the child's routine. This was undoubtedly due to a virtually wholesale commitment to the principle of 'less eligibility' (Radzinowicz and Hood, 1990). Nonetheless, by 1866 there were 65 reformatory schools accommodating up to 5000 young offenders (McConville, 1981, p. 338).

Early 'probation'

Although a 'formal' system of probation was not established until the early part of the twentieth century, its roots are to be found in nineteenth-century practices (Raynor and Vanstone, 2002). It has already been noted that the 1820s saw the emergence of diversionary reformative practices with juveniles. Later in the century, this practice was extended to some adult offenders. Edward Cox, recorder of Portsmouth, began the practice of releasing 'suitable' offenders on recognizance, a strategy which became the basis of the Summary Jurisdiction Act of 1879. This Act permitted a conditional discharge on the giving of sureties for good behaviour or reappearance for sentence. Cox believed that this was suitable for unpractised offenders in that it offered them 'a chance of redemption under the most favourable circumstances' (Cox, 1877, quoted in Radzinowicz and Hood, 1990, p. 634).

A parallel development sprung from the 'rescue' work of the Church of England Temperance Society (CETS), established in 1862 with a view to tackling the 'vice' of drunkenness and the reformation of the intemperate. Although not originally focused on offenders, the work of the CETS was from the mid-1870s extended to this group, a development which was reportedly prompted by a letter from Frederic Rainer to his friend and co-founder of the CETS, Canon Ellison (Vanstone, 2004). As Radzinowicz and Hood (1990, pp. 61–64) have explained, in the latter half of the twentieth century, crime was commonly attributed to excessive drinking and a succession of public inquiries were persuaded of a causal connection between alcohol and offending. At the turn of the century, they observe, between half and three quarters of crime was still being attributed to drunkenness, despite an acknowledgement that the latter was in decline. It was in this context that the extension of the CETS's work to offenders is best understood.

Thus began the practice of attaching missionaries to the criminal courts – initially in London but later elsewhere – with a view to 'saving the souls' of those habitual drunkards released to their care and oversight (McWilliams, 1983). Between 1880 and 1894 the number of full-time missionaries working in the courts increased from eight to seventy (Radzinowicz and Hood, 1990, p. 641). As Jarvis has observed, the police court missionaries were pioneers in the sense that 'they developed the concept of social workers working in and for the courts; they [also] provided the rudiments of the techniques of individual concern for and a personal relationship with, offenders in the open' (1972, p. 9). By the late 1870s the CETS had extended its focus to include discharged prisoners who were offered help with accommodation, employment and the like, as well as being invited to sign a pledge of abstinence on release (Ayscough, 1923, cited in Vanstone, 2004, p. 8).

Another influential late nineteenth-century figure in probation's 'pre-history' is Howard Vincent, a lawyer and former Director of

Criminal Investigation at Scotland Yard. In the 1880s, Vincent became the most persistent promoter of the cause to offer first offenders, both juvenile and adult, a chance of reform without recourse to the contaminating prison environment. In 1886, having visited Boston, Massachusetts, where an early system of probation was in operation (Bochel, 1976; Vanstone, 2004), Vincent introduced the Probation of First Offenders Bill. This Bill proposed that any person without previous convictions be released on probation, subject to a form of police supervision modelled on the ticket-of-leave system. In a letter to *The Times*, Vincent explained that his proposed scheme was 'calculated to save hundreds from a habitual life of crime, to give back to the State many an honest citizen, and to save the pockets of the taxpayer' (26 July 1886, quoted in Vanstone, 2004, p. 17). Although the resulting Act was a drastically amended version of Vincent's Bill, devoid of any formal supervision component,[14] it did constitute evidence of a growing willingness to substitute consignment to a harsh prison regime with an opportunity to reform – albeit that this option was restricted to those offenders deemed to deserve a 'second chance'.

A new era? The Gladstone Committee Report

By way of a conclusion, we return to the prison context and the publication, in 1895, of the Gladstone Committee Report (Departmental Committee on Prisons, 1895).[15] From the perspective of the present chapter, this is a significant document by virtue of its unambiguous support for reform as an aim of penal practice, albeit alongside a continued commitment to deterrence. In contrast to the Carnarvon Committee some thirty years earlier, the Gladstone Committee concludes that 'prison treatment should have as its primary and concurrent objects, deterrence, and reformation' (1895, para. 47). Elaborating on this position, the Report states:

> We think that [...] prison discipline should be more effectually designed to maintain, stimulate or awaken the higher susceptibilities of prisoners, to develop their moral instincts, to train them in orderly and industrial habits, and whenever possible to turn them out of prison better men and women, both physically and morally, than when they came in. (1895, para. 25)

The Report is also significant in that it conveys, in places, a view about the corrigibility of offenders which is rather more optimistic than we might expect from such a document produced at the close of the nineteenth century. That is, it suggests that the majority of offenders – not just young or first offenders – might be amenable to reform:

> There are but few prisoners other than those who are in a hopeless state through physical or mental deficiencies who are irreclaimable. *Even in the case of habitual criminals* there appears to come a time when repeated imprisonments or the gradual awakening of better feelings wean them from habitual crime. (1895, para. 33; emphasis added)

As for the *model* of reform which the Committee envisaged, a clear focus on morality is evident: there are throughout the Report numerous references to the desirable objective of 'moral improvement', and as such it clearly belongs within the reformative tradition described in this chapter. There is, however, relatively little mention of religion as a vehicle of reform, and religious instruction is not as prominent in the Report as Howard might have liked. Instead, the Gladstone Committee puts forward a model of reform in which the individual 'humanising' efforts and influences of prison staff and communications with 'respectable' family members and friends are given precedence over the sermon and the prison chaplain. The potentially reformative influences of education, exercise and a return to productive labour, too, are emphasised.

Beyond these general prescriptions, the Committee does not spell out a precise 'correctional' regime. However, far from being an oversight, this is in fact quite a telling aspect of the Committee's conclusions. 'The difficulty of laying down principles of treatment', the Report considers, 'is greatly enhanced by the fact that while sentences may roughly speaking be the measure of particular offences, *they are not the measure of the character of the offenders*' (para. 20; emphasis added). In this statement, as Garland (1985) has cogently argued, is an implicit rejection of a longstanding commitment to the formula of uniformity in penal treatment, in favour of a much more *individualised* ('elastic') approach, with the classification of 'like' prisoners as an essential starting point. First offenders, in particular, are singled out for 'special' treatment in new, centrally managed institutions inspired by the purported successes of the reformatory system, to cater for a broader age range of 17–21.[16] Reflecting the Committee's concern to tailor penal treatments to particular classes of offender, recidivist criminals, habitual drunkards and the 'weak-minded' are considered to be more suited to their own, specialist, institutions.

Clearly we have done little more here than provide some highlights from the Gladstone Committee Report, focusing on its stance in relation to reform as a penal objective and practice. We should, however, note that the significance of this particular document is hotly disputed by penal historians. On the one hand, many regard its publication as a pivotal moment in the history of the 'reformative ideal' (e.g. Garland, 1985; Cavadino, 1995; Muncie, 2001). For example, in Garland's view, the Report represents a significant rhetorical departure from a Victorian penal rationality which afforded the objective of reform no more than a subsidiary place. In contrast, the argument has been put that 'far from ending a Victorian pattern of punishment, the Committee underpinned and reinforced it for at least another half century' (Blom-Cooper and McConville, 1995, p. 1; see also Forsythe, 1991; McConville, 1995b). Some penal historians have also pointed out that the actual impact of the Gladstone Committee Report on penal policy and practice was far from revolutionary, at least in the short term, and this conclusion is probably fair (Radzinowicz and Hood, 1990; Forsythe, 1991). Nonetheless, taking a slightly longer perspective and, importantly, looking beyond the prison context, it is possible to view the Gladstone Committee Report as a significant step in the direction of a penal system, and a penal rationality, which would look rather different from the one which became established in the nineteenth century.

This is essentially the thesis put forward by Garland (1985). For Garland, it is principally the introduction of the character and history of the individual offender into the calculus of penal thought and practice, along with a new 'official' confidence in the possibility of reform – both prominent features in the Gladstone Committee Report – which signifies the inauguration of a distinctly modern penal complex, the essential elements of which, he argues, were in place by 1914. These elements would include the establishment of a system of probation; a new sanction of Borstal training; detention in reformatories for the inebriate and a separate system of detention for the mentally defective; as well as various forms of licensed supervision (see also Radzinowicz and Hood, 1990, pp. 573–99). In this new system of penality, characterised by Garland as 'penal-welfare', punishments would become more differentiated and less prison-centred; more attuned to different categories or classes of offender, rather than to different offences *per se*; and oriented, to a much greater extent than at any time in the past, towards the objective of reform – and, later, 'rehabilitation'. It is to this recalibrated penal system, and the rise (and fall) of a profound rehabilitative optimism, that we turn our attention in Chapter 4.

4
The Rehabilitative Ideal: Advance and Temporary Retreat

The Gladstone Committee successfully and decisively inserted rehabilitation into official discourses concerning the purposes of the penal system, but implementation of such ideas was to be a slow process spanning many decades. This chapter describes how rehabilitation, originally an aim in search of a method, became largely identified during the middle decades of the twentieth century with the aims and methods of the emerging social work semi-profession. It goes on to describe how rehabilitation suffered a major setback as the century moved into its final quarter, when the industrial societies of the West became sceptical about the cost and effects of the Welfare States which had been built with such effort and commitment only a few decades before. A modernist optimism about the efficacy of a claimed scientific and professional approach gave way to a late-modern doctrine that 'nothing works', and for a time it looked as if both the rise and the fall of rehabilitation would be encompassed within less than a century of penal history.

Rehabilitation in search of a method

Radzinowicz and Hood (1990) document the slow process of modest experiment which began to bring some of the Gladstone Committee's ideas into practice. Penal reformers were particularly concerned about young men who were beginning to offend but who were not yet habitual or incorrigible criminals, and for whom there was, as the Prevention of Crime Act of 1908 put it, a 'reasonable probability that [he] will abstain from crime and lead a useful and industrious life' (s. 5). The Gladstone Committee had been impressed by the Elmira reformatory in New York, which was also much admired by some influential

European criminologists such as Raymond Saleilles (Radzinowicz and Hood, 1990). British penal administrators chose to advance more cautiously, and devised a regime for young offenders based on strict discipline, education, training, exercise and constant activity. Small-scale experiments began in 1900 at Bedford prison, and were extended to other prisons at Borstal (which eventually lent its name to the new type of regime) and Dartmoor. The Prevention of Crime Act 1908 introduced a new flexible sentence of detention in a 'Borstal Institution' for young offenders aged 16–21, for a period of one to three years, with the option of early release on licence for good progress.

In spite of some concern that these young people, if judged suitable for Borstal, would in many cases serve longer sentences than was normal for the offences they had committed, the new measure proved reasonably popular, and was an early example of rehabilitative aims taking precedence over principles of proportionality. (The same Act introduced preventive detention, another disproportionately long sentence designed to prevent offending by habitual criminals through what is now known as incapacitation.) The regime, based largely on foreign models such as Elmira, involved heavy doses of early rising (5.30 a.m.), drill, exercise and gymnastics, with some elementary education, plenty of manual labour and morally improving sermons. Trainees who responded well to this were promoted through a series of grades marked by different coloured ties, with increasing privileges and responsibilities culminating in supervised release. A Borstal Association was formed to provide post-release supervision through voluntary associates, and high success rates were claimed, although these probably resulted mainly from the fact that many trainees did not have particularly long criminal records at the time of sentence.

The Borstal regime was to persist in recognisable form for most of the century, to be immortalised in literature such as Brendan Behan's *Borstal Boy* (1958) recording his experiences in the 1940s, and Alan Sillitoe's *Loneliness of the Long-distance Runner* (1959). However, this continuity masked important changes both in the way the institutions were used, and in the ideas about young offenders and their rehabilitation which informed them. Their niche in the penal marketplace moved gradually 'up-tariff', especially following the introduction of Detention Centres in 1948 for those young offenders who did not require Borstal, and also as a result of the increasing use of non-custodial penalties such as probation. Young offenders with longer records were more likely to re-offend on release, and by the 1970s the Borstal institutions had the highest reconviction rates in the system. Meanwhile, the change in ideas was so marked that by the beginning of the 1960s we find the Governor of Hewell Grange Borstal writing an article in the *Prison Service Journal* which is not about discipline or exercise but about 'Casework in Borstal' (Roberton, 1961). The language of rehabilitation had changed, and new ideas, methods and professions had made their presence felt. What was this 'casework', and where had it come from?

The dominance of social casework

The main source for the emergence of social casework into the field of rehabilitation and corrections was, of course, the probation service. The 1907 Probation of Offenders Act was another product of that extraordinary first decade of the twentieth century which saw so much progressive legislation in the field of social policy. It created a new group of specialists within the criminal justice system, the probation officers, and it looked to them to provide firm but sympathetic guidance to young offenders who might, as a result, avoid graduating to prison and a life of crime. In the early years, the methods and techniques of probation supervision were based on those previously developed within the police court missions, but some officers quickly aspired to a more secular, professional and 'scientific' standing. Vanstone (2004) has called this the movement 'from Awakening the Conscience to Providing Insight' (p. 67), and documents the astonishing variety of social and psychological theories which caught the attention (usually briefly) of the early probation officers. From eugenics to the dangers of masturbation to the early psychology of Cyril Burt (1925), probation officers were eager for any knowledge which carried the stamp of more established expertise, although many tempered this with their own common sense and life experience.

By the time the Probation Training Board was set up following a Departmental Committee Report in 1936, the most relevant disciplines for understanding and accounting for delinquent behaviour were generally held to be psychology and psychiatry (Rose, 1985). Clearly these should inform the new training; on the other hand, probation officers could not be expected to undergo the long and expensive training required for these professions, and in many cases would not have been qualified to undertake it anyway. Fortunately this dilemma had already been faced by another emerging group of welfare specialists, the social workers, particularly in the United States. In that country, probation work and other work with young offenders were already seen as a significant branch of social work, and shared in the emerging mode of practice known as 'casework'. Vanstone dates the adoption of 'casework' by the British probation service to the 1930s: as so often in the field of rehabilitation, the cautious and piecemeal process of public administration in Britain followed where other countries had led, and claims of pioneer status were seldom justified.

The term 'casework' originated in the Charity Organisation Societies (COS) of the nineteenth century. COS workers would make individual assessments of applicants and their families for assistance, partly to distinguish deserving from undeserving applicants and to ensure that any sums granted in assistance would be wisely spent. Aiming to fill the gap left by harsh and inadequate systems of poor relief, they were very conscious of the need to avoid dependency, and they aimed to assess the social circumstances and moral character of their applicants (Mowat, 1961). (In this they resembled the early Almoners, whose task was to ascertain which hospital patients were genuinely in need of charitable assistance and which could actually afford to pay, or had relations who could.) 'Casework' was, originally, simply 'work on cases': each applicant became a case and generated a case file. The original social workers worked for these societies, assessing applicants for charitable welfare assistance and supervising their use of it: their method was 'social casework', but as yet this did not imply any particular specialist knowledge base.

This began to change through the pioneering work of Mary Richmond, a distinguished American charity organisation caseworker whose very influential book *Social Diagnosis* (1917) argues for a scientific approach informed by the social sciences and by an understanding of the causes of poverty and other social problems. However, in the decade following the First World War, as the number of social workers increased and more of them were employed by various public services, the search for a recognised and distinctive method and the accompanying professional recognition grew stronger. Like other workers in human services in America, social workers were eager to learn more about the new psychodynamic psychology which was emerging from Europe and offered a new understanding of personality development, maladjustment, neurosis and many other problems which social workers regularly encountered. New ideas based on psychoanalysis and 'depth psychology' had already found their way to America and were shaping the 'mental hygiene' movement: as early as 1909 Freud gave lectures on psychoanalysis at Clark University in Massachusetts.

The social work educator, psychiatric social worker and therapist Jessie Taft (1926) provides both a good account and an example of the development of social work at this time. After postgraduate education at the University of Chicago and a PhD in philosophy with a dissertation on *The Woman Movement from the Point of View of Social Consciousness* (Taft, 1913), she found work in the rapidly developing 'mental hygiene' movement, including experience with children in a farm school, women offenders in a State Reformatory and people with mental health problems looked after by the first 'mental hygiene clinic' in New York. Before long she was working with children in Philadelphia and teaching social work at the University of Pennsylvania, with her friend Virginia Robinson who had joined the staff there towards the end of the war. The social work world in which these two able and committed young women were making their way was a mixture of the old charities and settlements, Mental Hygiene Committees, over-stretched clinics for adults and children with mental or behavioural problems, penal institutions which sometimes aspired to help rather than punish, and less familiar components such as the Laboratory of Social Hygiene and 'eugenics fieldworkers' who studied the family histories and relationships of needy rural families. Both Taft and Robinson had briefly worked as 'eugenics fieldworkers', and Robinson had also been active in the women's suffrage movement in Kentucky (Robinson, 1962). Taft even described 'eugenics fieldwork' as a precursor of psychiatric social work. (Both women went on to long and distinguished careers working together in Philadelphia; Taft became a pupil of the psychoanalyst Otto Rank, and completed a biography of him [Taft, 1958] before her death in 1960.)

In this strongly reformist and progressive period for social work, no longer charity but not yet science, there was a great hunger for some theoretical basis. Early lectures on 'mental hygiene' offered to social workers proved so attractive that people climbed through windows to get seats: 'Atlantic City in 1919 was a landslide for mental hygiene: the [National] Conference [of Social Work] was swept off its feet' (Taft, 1926). Although social workers were hardly in a position to offer their 'clients' years of analysis based on such techniques as free association or the interpretation of dreams, they quickly came to see their 'casework' as essentially a therapeutic process drawing on psychodynamic theory to help 'clients' to understand their own problems and behaviour. Mary Richmond's emphasis on social processes was replaced by the concept of the 'psychosocial' in which the inner world of the 'client' interacted with the environment to produce healthy or unhealthy 'social functioning'. Jessie Taft herself described the excitement generated by these new ways of thinking: 'It is not often that one has the experience of witnessing the birth of an epoch-making movement. One hundred years from now the story of the beginning of mental hygiene will be a vital chapter in the history of our culture' (Taft, 1926, p. 56). Different theories of social casework quickly emerged (see Roberts and Nee, 1970), sometimes echoing the acrimonious splits which were already afflicting the psychoanalytic movement, but the strongest international influence (certainly as far as Britain was concerned) was wielded by the 'diagnostic school', which eventually produced Florence Hollis's classic textbook *Casework: A Psychosocial Therapy* (1964). By then the transformation was largely complete: the leaders, educators and textbook writers of the social work profession saw it not as the efficient administration of welfare but as a form of therapy, not quite psychotherapy but almost. (One can only speculate how different the history and impact of social work might have been if they had instead chosen the road of social science and informed social activism, or of empirical social psychology rather than the more speculative constructs of Freud and the psychoanalysts.)

In Britain some training of social workers had been established towards the end of the nineteenth century through social science courses initially designed for COS and Settlement workers.[1] Some University involvement can be traced as far back as 1904 when a 'School of Social Science' was set up in Liverpool by the University, the Victoria Settlement for Women and the Liverpool Central Relief and Charity Organisation Society (Macadam, 1925). Other Universities in major cities soon followed: Birmingham, Bristol, Leeds, Edinburgh (in the School of Cookery and Domestic Economy), Glasgow and eventually London. Such courses were initially and usually offered through the extra-mural or adult education departments of universities rather than as core subjects. The mixture of social science, economics, law and public administration was devised by experienced social workers, some of whom also wrote textbooks in the tradition of social administration rather than depth psychology. (These early authors included Elizabeth Macadam, the lifelong friend of the social reformer and pioneering woman MP Eleanor Rathbone, and also Clement Attlee, the future Labour Prime Minister [Attlee, 1920; Macadam, 1925, 1945]. The courses set up to train social workers in London developed eventually into the London School of Economic and Political Science, which no longer trains social workers.) However, by the time training for probation officers was set up in the late 1930s the psychodynamic influence was more evident, and the dominance of casework was consolidated after the Second World War.

In the late 1940s and 1950s many British social workers visited the United States and participated in training and professional activities there before returning to Britain and moving into senior posts and training positions on the basis of their American experience. A number of textbooks appeared (such as Biestek 1961 and Timms, 1962), together with new training practices such as a form of supervision which seemed as much concerned with the inner psychodynamics of the supervised social workers as with what they were actually doing (Towle, 1954; Pettes, 1967; Young, 1967). Publishers even launched specialist series of social work texts: Allen and Unwin published the National Institute for Social Work Training Series, and Routledge and Kegan Paul launched the Library of Social Work, including such works as Raymond (later Lord) Plant on *Social and Moral Theory in Casework* (1970), and Isca Salzberger-Wittenberg on *Psycho-Analytic Insight and Relationships: A Kleinian Approach* (1970). Probation training was increasingly integrated with social work training as, in addition to the Home Office course, probation 'streams' were opened in several Universities' social work courses. Briefly, if the treatment model (see Chapter 2) provided the rationale for rehabilitative work with offenders, psychosocial casework aimed to provide the methods by which offenders could gain insight into themselves. Along with the quest for specific knowledge and technique and the accompanying professional status developed an aspiration towards 'generic' social work. This was the idea that all social workers, regardless of setting, were essentially engaged in the same tasks and using the same skills, and could therefore be trained together. Social work texts of the 1950s tended consistently to take this position, and probation work was routinely counted among the social work occupations which shared this generic base.

A small volume in the possession of one of the authors was published by the Association of Psychiatric Social Workers in 1957 and illustrates a number of tensions arising from the attempt to establish a profession (Goldberg *et al.*, 1957). Entitled *The Boundaries of Casework* (which already suggests some anxieties about whether too much was being claimed), it gathers together the papers given on a five-day residential refresher course for psychiatric social workers. This was one of a series going back to 1952 (when they discussed 'the application of casework skills in a variety of settings'). The contributors included several psychiatric social workers, a psychiatrist, a sociologist (Professor Simey from Liverpool) and a criminologist/youth worker (J.B. Mays). Among the psychiatric social workers are some who went on to become leading authors and trainers in their field, such as Kay McDougall, Jean Snelling,

Margaret Ferard and Noel Hunnybun. Probation is assumed to be within the compass of generic social work, but it is not known if any probation officers attended.

The speakers refer frequently to events in America: one recalls with pleasure a conference on how to implement generic fieldwork for training purposes, while other speakers raise some concerns about where social work is going: for example, the sociologist points out that the gaps in social work theory have been filled by 'copious drafts of "depth" psychology of one school or another', and that in America 'the client's emotional disposition has become more and more the object of attention, and his environment less and less a matter of serious concern'. As a result, 'the American caseworker has also tended to lose touch with the formulation and execution of social policy'. Moreover, 'the emotional basis of the relationship between teacher and student in the "supervision" situation has...been termed a "cult"' (Simey, 1957, pp. 61–62). His fellow social scientist described an experimental youth club in which 'the therapy we offer is social rather than psycho-analytical' (Mays, 1957, p. 73). One cannot help wondering how all this went down with the aspiring psychosocial therapists, and whether these speakers were invited again. The insecurity of the developing profession is hinted at by another speaker who argues that 'our work approaches most nearly to that of psychotherapy...under certain conditions social caseworkers can practise a form of psychotherapy...social casework is not an attenuated form of psychotherapy but an entity in its own right' (Davies, 1957, p. 36).

Casework and 'Authority'

In keeping with the atmosphere of post-war penal modernisation, expectations of the probation service were high and evaluations (without much evidence at that time) were strikingly positive. In 1952 Max Grunhut (a German lawyer and criminologist who, like Hermann Mannheim and Leon Radzinowicz, had escaped from the Nazi regime before the war and helped to found the academic discipline of criminology in Britain) wrote: 'Probation is the great contribution of Britain and the USA to the treatment of offenders. Its strength is due to a combination of two things, conditional suspension of punishment, and personal care and supervision by a court welfare officer. With the growing use of probation, social case work has been introduced into the administration of criminal justice...' (Grunhut, 1952, p. 168). In a similar vein, Manuel Lopez-Rey, head of the United Nations Social

Defence Section, wrote: 'If I were asked which, among the modern methods for the treatment of offenders is the most promising, without hesitation I would say: Probation' (Lopez-Rey, 1957, p. 346). (The Howard Journal, in which both these comments appeared, in those days bore the subtitle 'A review of modern methods for the prevention and treatment of crime and juvenile delinquency'.) In 1958 Leon Radzinowicz wrote: 'If I were asked what was the most significant contribution made by this country to the new penological theory and practice which struck root in the twentieth century... my answer would be probation' (Radzinowicz, 1958, p. x). These were the opening words of his preface to *The Results of Probation*, one of the first British studies of probation outcomes, which found reconviction rates to be low but used no comparison group. Nevertheless such broadly favourable views set the tone for approximately the next twenty years.

Acceptance of the casework message was not without some difficulties for probation officers. The therapeutic relationship was presented in the literature as essentially voluntary rather than coerced, while the probation officer might be obliged by his role to prosecute his (usually in those days his) 'client' for not engaging in it. Leading thinkers in probation were committed to the Service's position as an important branch of social work, and they shared the professional aspirations of their colleagues in other branches, even if the rank and file of the probation service retained, in many cases, a degree of scepticism about psychodynamic models of offending and social work. As one concerned social work writer put it in the early 1960s, 'a number of men officers and some women officers carry out their probation work in an authoritarian, directive way, paying little heed to such things as maintaining a non-judgmental attitude and self-determination for the client, and apparently work successfully' (Eshelby, 1962, p. 126). The problem was often articulated as a conflict between 'care' and 'control': how could these be reconciled? A few social work theorists argued that probation could not be true social work because of the absence of true voluntarism; as the distinguished social work educator W.J. Reid later argued, caseworkers in correctional settings might need to take authoritative action against the client's wishes, but 'we should understand that such actions, while perhaps necessary, do not constitute casework treatment' (Reid and Epstein, 1972, p. 215). However, rescue was on its way, and once again the mental health professions made their contribution.

In 1964 an article appeared in the British Journal of Criminology by A.W. Hunt, Principal Probation Officer of Southampton, entitled

'Enforcement in Probation Casework'. In this paper he argued that enforcement is required because immature probation clients do not fully understand what is in their own interests, but that this does not infringe the essentially voluntary nature of a therapeutic relationship because later on, when more able to make a mature judgment, clients would come to see that enforcement had been in their own interests and would thereby give a kind of retrospective consent. In the style of social work writing of the time, this is illustrated with a few selected case studies rather than any broader or more structured form of research. One of these case studies concerns a boy sent to Borstal, probably on his probation officer's recommendation, who later writes to thank the officer: 'really what I needed when I was little was a good hiding but I always knew I would not get one, but this is just as good as one although it has taken a long time to bring me to my senses' (Hunt, 1964, p. 246). Hunt also quotes with approval the view of Clare Winnicott, a psychiatric social worker, about the unconscious motivation behind criminal offences: 'symptoms of this kind of illness [*sic*] are unconsciously designed to bring authority into the picture' (Winnicott, 1962, p. 181). So not only could probation officers rely on the prospect of retrospective consent to make their practice consistent with social work principles of self-determination, but perhaps their clients were really volunteers all along, albeit unconsciously. Hunt sums up in this way: 'The enforced relationship and casework are not mutually exclusive. Indeed, in many respects the probation casework process is enriched by enforcement...' (Hunt, 1964, p. 251).

These theories were eventually restated at length in a very influential book by two of the leading thinkers in probation practice and training at the time. Foren and Bailey's book *Authority in Social Casework* (1968) became the core text for those who saw probation as a major branch of social work. The authors argued a convincing case that most social work, in most settings, contained elements of authority (few people, then or now, *volunteer* to be helped by social workers) and to their credit they even incorporated an appendix on research. However, this nod in the direction of criminology did not seem to lead to any questioning of the book's basic pathologising assumption that offenders were 'immature'. Shortly afterwards, the probation service came close to amalgamation with other branches of social work in Local Authority Social Services Departments. This was recommended by the Seebohm Report in 1968, and seen by leading social workers as the logical organisational consequence of the generic nature of social work and, by then, social work training (Hall, 1976). Only a campaign by the courts and by some probation officers prevented amalgamation in England and Wales. (In Scotland, probation services were merged into Social Work Departments, and eventually became their Criminal Justice Social Work teams when specialisation re-emerged in the 1980s.)

The decision not to amalgamate in England and Wales reflected the fact that many probation officers saw themselves as different from local authority 'social workers' (the term used in the new Social Services Departments to cover former Child Care Officers, Mental Welfare Officers, Welfare Officers and in some cases Educational Welfare Officers). However, this difference was largely seen to exist in their particular relationship with the courts. Most officers at this time saw their professional affinities and their methods as belonging within social work (Vanstone, 2004), and social casework as their main method of work. The early 1970s represented probably the high point of social casework's dominance of the rehabilitative agenda. However, challenges to this position were by then just around the corner, and the dominant orthodoxy which defined probation work as social casework in a context of authority was vulnerable both to questions about the efficacy of rehabilitation and to wider questions about the casework enterprise itself.

Doubts emerge about 'casework'

The rise of casework theory to dominance among social welfare professionals had not been without its critics: for example the American sociologist C. Wright Mills drew early attention to the possibility that a professional doctrine (or, as he defined it, a professional ideology) which attributed social problems to individual malfunctions would stand in the way of policies designed to address structural inequalities and promote opportunities (Mills, 1943). (As we have seen, similar comments were later made by Simey to the psychiatric social workers' refresher course.) In addition to the early emergence of a critique based on social theory, there was also some early pressure towards empirical testing: as early as 1931 Dr Richard C. Cabot's presidential address to the Association of Social Workers called for evaluative research to support their claims, and an experiment was soon designed which embodied some of the latest thinking about evaluation, in particular a random allocation design in which subjects were randomly assigned to a 'treatment' or a 'non-treatment' condition. This was the Cambridge-Somerville Youth Study, eventually published as *An Experiment in the Prevention of Delinquency* (Powers and Witmer, 1951).

It is not accidental that the chosen focus of this early evaluation was the prevention of juvenile delinquency. Social work has always been closely involved in issues around criminality, particularly the criminality of young people. Social workers may have wanted to see themselves as therapists promoting development and self-realisation in voluntary 'clients', but when they worked for public authorities, those who paid them were more likely to expect that they would contribute to effective public administration and social control. Juvenile delinquency was social work territory well before most probation officers saw themselves as social workers. It is also interesting that this major study was set up in progressive Massachusetts, where so many early developments in charity organisation and probation had taken place. This was not a hostile evaluation carried out in an anti-welfare atmosphere.

The experiment was based on a sample of 650 adolescent boys. Half of them, chosen at random, were allocated the support of a social worker over six years (1939–45), and the essence of the experiment lay in comparing their behaviour with the other half who did not have social workers. The result was that those with social workers offended no less than the others (in fact slightly but not significantly more: 96 boys in the experimental group committed 264 offences, while 92 members of the control group committed 218). Oddly, the social workers thought they had substantially helped about two-thirds of the boys under their supervision, and more than half of the boys who had social workers believed they had been helped, but the results suggested no positive behavioural impact. However, the most worrying outcomes were revealed in a follow-up by Joan McCord thirty years later (McCord, 1978). In a remarkable study she was able to trace 95 per cent of the 506 original subjects who were in the experiment after 1942 and found that the experimental group had become, over time, significantly worse off than the control group on a variety of indicators. These included crime, occupational status, alcohol misuse, mental illness, stress-related illnesses including high blood pressure and heart trouble, and life expectancy. These astonishing results are hardly ever discussed in the social work literature.

Early discussion of the initial outcomes of the Cambridge-Somerville experiment questioned whether some of the social work had been of the best quality, and attempts were made to control this in other studies by ensuring that all the social workers were appropriately quali- fied. However, results tended to be no better: for example, in another high-profile random allocation study targeting 'pre-delinquent' girls, the 'Vocational High' experiment, a variety of outcome measures showed no significant benefit to the experimental group (Meyer *et al.*, 1965). Introducing the published report of this study, Leonard Cottrell of the Russell Sage Foundation famously asked the question 'Is social work on the wrong track?' (ibid., pp. 3–4). The challenge to the profession was posed in the following terms: 'the issue is whether or not the social work profession can continue to maintain that the individual casework, clinical approach is its central method for dealing with the kinds of problems presented by the population dealt with in this experiment'. There were also indications that some of the girls resented being picked out as 'pre-delinquent'.

Discussion of these experiments and reactions to them were found largely in the American social work journals. For example Grey and Dermody (1972), reviewing these and other studies showing no benefit from casework (such as the Chemung County evaluation of services to multi-problem families [Wallace, 1967]), made a number of methodological points. They argued that more realistic studies could be done using people who had actually asked or been referred for a service, and that more needed to be done to ensure and standardise the quality of work done with the experimental groups. However, in so far as the experiments represented an evaluation of social work in situations where social workers expected to have an impact, the results were not encouraging. This trend was confirmed by a systematic review carried out by Joel Fischer, who aimed to cover all evaluative studies of casework which met basic criteria of methodological adequacy, identified seventeen such studies and found that overall they did not support the effectiveness of casework methods as practised in contemporary social work (Fischer, 1973, 1976). His overall conclusion was that 'The bulk of practitioners in an entire profession appear, at worst, to be practicing in ways that are not helpful or even detrimental to their clients, and, at best, operating without a shred of empirical evidence validating their efforts' (Fischer, 1976, p. 140). In the book in which he wrote up these findings at length, he questioned the theoretical basis of psychodynamic casework as well as its practical efficacy, and concluded that research to date had supported only three promising directions of development. These were the use of structured methods, including time-limited brief contact; the use of behavioural methods to change behaviour, and the development and maintenance of relationships characterised by the 'core therapeutic conditions' of empathy, warmth and genuineness (Fischer, 1976).

The evidence for structured and time-limited methods came partly from a controlled comparison between brief and extended intervention with voluntary clients, in which structured contact over a few months produced better results than indefinite contact (Reid and Shyne, 1968). Reid went on to develop and apply these findings in a new framework for social work practice which he called task-centered casework, based on a clarity about agreed objectives or 'target problem', time limits and agreed tasks for both social worker and client (Reid and Epstein, 1972). A number of studies pointing to the efficacy of 'core therapeutic conditions' based originally on the work of Carl Rogers had been carried out and helpfully summarised by Truax and Carkhuff (1967): in particular they showed that these could be reliably identified from observation of helpers' behaviour, and could be increased by appropriate training. Theoretical orientation was less important as a determinant of effectiveness than helpers' actual behaviour, and effective therapists tended to become more effective over time, whilst ineffective therapists tended to become worse, but their belief in their own effectiveness often increased over time. They argued, and Fischer agreed, that one probable reason for findings of 'no average benefit' in evaluation studies was that no attempt was made to distinguish between effective and ineffective social workers: the methods used were identified under the broad heading of 'casework' rather than identifying what workers actually did, so that any benefits produced by effective work might be cancelled out by deterioration associated with ineffective work, and this would not be evident to the evaluators. This emphasis on helpers' actual behaviour rather than their beliefs, attitudes, feelings or insight was a challenge to the way social workers were trained, and to the practice of assessing competence by discussing cases rather than observing casework; however, the increasing emphasis on observable behaviour was also a feature of Fischer's second group of promising methods, and he went on to develop the practical implications of these findings in further books (Fischer and Gochros, 1975; Fischer, 1978). Others concentrated on the 'core therapeutic conditions' and, particularly in the area of counselling, a skills-based literature developed, eventually feeding back into social work through texts such as Egan's *The Skilled Helper* (1986).

Other American commentators were drawing similar conclusions about the general effectiveness and evidence-base of casework as taught and practised at that time (for example, Mullen and Dumpson, 1972; Wodarski and Bagarozzi, 1979). Reactions in social work ranged all the way from extreme defensiveness, to the adaptation and incorporation of new ideas and methods in training courses. In Britain the experience was rather different: the new empirical material from the USA was slow to penetrate into training courses. In the 1960s and early 1970s these were based mainly in Departments of Social Administration with a limited empirical tradition, and were often run by people who believed their main task was to socialise trainees into the accepted thinking and practices of the profession. (When more critical material began to be introduced by lecturers, it was sometimes described by the more orthodox as 'educating for uncertainty' or 'teaching about social work instead of teaching social work'.) However, another kind of critique was emerging in Britain, and had as much impact on probation work as on other areas of practice.

Social work as ideology

In America the most telling criticisms of social work tended to come from evaluation experiments, using quantitative methods based on the examples of medical or psychological research. In Britain more of the critical literature tended (like Mills in the USA) to be based on social theory, and sometimes on social research methods. In 1959 Wootton's *Social Science and Social Pathology* reviewed criminological writings on offending, and criticised both psychiatrists and (particularly) social workers for confusing social and medical/psychological phenomena. Her chapter on 'Contemporary attitudes in social work' mocks the more grandiose claims of casework and includes the memorable statement 'It might well be thought that the social worker's best, indeed perhaps her only, chance of achieving aims at once so intimate and so ambitious would be to marry her client' (p. 273). Some years later, Mayer and Timms (1970) studied the views of clients of the Family Welfare Association, and found many of them to be confused by the casework service they had received. They did not understand why the social workers were so interested in their recollections of childhood, nor why they were so uninterested in the immediate reason for them being there, which was often some financial or other material crisis. They were also puzzled by the social workers' reluctance to give advice. (Here we can recognise the psychodynamic theory of the 'presenting problem' which masks the real, underlying but unconscious reason for seeking help.) Mayer and Timms showed a clear mismatch between clients' and workers' perceptions of the problem, but ended with the rather surprising suggestion that prospective clients should be offered a short training in what to expect from casework before embarking on it.

Others were more likely to suggest that the methods needed to be modified: for example Sinfield (1969) produced a Fabian Society pamphlet *Which Way for Social Work?* which argued that if social workers continued to use a psychodynamically based individual case-work model they would not be well placed or equipped to play their part in social programmes which attributed many individual problems to poverty and powerlessness, and aimed to address these through changing the distribution of wealth and opportunity. This issue was also raised by many new recruits to the profession in the early 1970s: influenced by New Left ideas in the Universities of the late 1960s and more sympathetic to a Marxian than a Freudian analysis of the problems of the poor, they developed a more structural analysis of social problems and an understanding of social work which was more about empowerment than about therapy (Bailey and Brake, 1975; Leonard, 1975; Corrigan and Leonard, 1978; Simpkin, 1979; Brake and Bailey, 1980). Similar points were made about community development and community work (Loney, 1983).

Although such ideas (usually known as 'radical social work') never became the mainstream of social work thinking (particularly in the Social Services Departments), they had (together with new anti-racist and feminist influences) a considerable influence on later developments such as anti-oppressive practice (Dominelli, 1988, 2002). While their socialist political prescription may be short of adherents in the New Labour age (Jordan, 2000), the tension they highlighted between explanations based on individual pathology and those based on social structures and influences has never gone away, and perhaps remains essentially unresolved within social work. Most importantly, they pointed to an ideological function of social work: in advocating and promoting individualised solutions to social ills, social work could serve the interests of powerful groups in society by diverting attention from the need for social remedies – the idea of social work as a social tranquillizer. Such arguments provided little guidance on what should be done for those suffering now and unable to create or even wait for massive social change; however, the notion that social work has an ideological function is important, and we will return to it when considering why certain ideas about the effectiveness of rehabilitation receive such different degrees of political support at different times.

Doubts about probation: The emerging orthodoxy that 'Nothing Works'

The territory of probation was at this time still regarded as part of the continent of social work, and questions about the efficacy of social work raised serious questions for probation, especially as so much of the social work research had actually concerned delinquency. Other movements within social work were also reflected in probation, for example the vogue for 'systems theories' (Pincus and Minahan, 1973) which were energetically taken up by the Central Council for Education and Training in Social Work (CCETSW).[2] The radical social workers' campaigning magazine *Case Con*, published by a collective during the 1970s, had its counterpart in the NAPO Members' Action Group and its newsletter (NAPO is the National Association of Probation Officers, the Trade Union and professional association for probation workers). A radical model of probation was proposed and elaborated by some of its members (for example in Walker and Beaumont, 1981). However, distinctive problems were also emerging from criminological research which questioned the effectiveness of all rehabilitative efforts, including probation.

The emergence of a consensus that 'nothing works' is usually traced to Robert Martinson's article of 1974 in the American periodical *The Public Interest*. This article, entitled 'What works? Questions and answers about prison reform' reported a summary of the headline results of a review of over two hundred studies of rehabilitative services to offenders in both custodial and community settings. There were some indications that the findings were seen as threatening by established interests: the review was initiated in 1996 by the New York State Governor's Special Committee on Criminal Offenders and was completed by 1970, but the State did not publish it, and did not allow the researchers to do so until disclosure had been forced by a court case. Martinson's overall conclusion is that the results 'give us very little reason to hope that we have in fact found a sure way of reducing recidivism through rehabilitation . . . education at its best, or . . . psychotherapy at its best, cannot overcome, or even appreciably reduce, the powerful tendency for offenders to continue in criminal behaviour' (p. 49). He attributes these failures partly to the theory of crime as a disease, and points out the normality of both crime and 'a very large proportion of offenders, criminals who are merely responding to the facts and conditions of our society'. The article ends by advocating decarceration for low-risk offenders and a research-based reappraisal of 'to what degree the prison has become an anachronism and can be replaced by more effective means of social control' (p. 50), but what was remembered was the broad conclusion about the ineffectiveness of rehabilitation, summed up in the slogan 'nothing works'.

The report behind the article, when eventually published in the following year (Lipton, Martinson and Wilks, 1975) did not exactly say this anywhere in its 735 pages. It incorporated a number of summaries of findings in relation to different treatment approaches and outcome measures, and in relation to recidivism it found mostly, but not entirely negative results. For example:

> To the degree that casework and individual counselling provided to offenders in the community is directed towards their immediate problems, it may be associated with reduction in recidivism rates. Unless this counselling leads to solution of problems such as housing, finances, jobs or illness which have high priority for offenders, it is unlikely to have any impact upon future criminal behaviour. (p. 572)

However, this and many other carefully balanced summaries were re-summarised by Martinson (1974) as '*With few and isolated exceptions, the rehabilitative efforts that have been reported so far have had no appreciable effect on recidivism*' (p. 25, italics in original).

Similar findings emerged from other research reviews: for example, Brody's review in Britain for the Home Office (Brody, 1976) concluded that 'any assumption that that different sentences, institutions or "treatments" are affecting offenders in different ways needs to be carefully reappraised' (p. 66). Although many of the studies included in these reviews must be subject to the same reservations as those included in Fischer's casework review which was discussed earlier in the context of social work (Fischer, 1976), namely that methods were often not well specified nor consistently delivered to adequate standards, the dominant response was to accept the verdict that nothing worked.

In Britain, additional confirmation of this seemed to be emerging from a substantial programme of research on probation. Around the same time as Radzinowicz's study, Wilkins had produced an outcome study of probation orders which had rather less encouraging results, mainly because he used a comparison group (Wilkins, 1958). The Home Office was then engaged throughout the 1960s in an ambitious programme of probation research aimed largely at developing empirical classifications of probationers and their problems, leading to such impressively detailed studies as Davies (1969). This was followed by an attempt to describe and classify what probation officers actually did about these problems (Davies, 1974), which included among its conclusions the recommendation that a properly controlled experimental study should test the effectiveness of probation officers' work. Such a study was indeed set up, based on random allocation of probation clients to normal or 'intensive' caseloads, to test the probation service's contention that more probation officer input would lead to better results (the 'IMPACT' study, reported by Folkard *et al.*, 1974, 1976).

Like much early research on the effects of reduced caseloads (but not all: see Adams, 1967), the results of this experiment were disappointing and, although they were not much discussed within the probation service, they had a significant impact on the research agenda. Probation officers in the experimental (intensive) group did not deliver any special intensive intervention designed to reduce offending: instead, left largely to their own devices, they mostly just did more of what they would normally do. The results were summed up as 'small non-significant differences in reconviction in favour of the control cases' (p. 15): in other words, higher 'doses' of probation were no more effective, and possibly less effective, than lower 'doses' (out of 475 male probationers in the experimental group, 38.1 per cent were reconvicted in one year, compared to 33.6 per cent of 425 male probationer 'controls'. Similarly, 44.8 per cent of 29 female 'experimental' proba-tioners were reconvicted, compared to 34.5 per cent of 29 female 'controls').

The only group which stood out as a significant exception to this general result was a fairly small number of offenders who combined a low level of 'criminal tendencies' (a variable based on probation officers' assessments, and not based simply on length of criminal record) with a high level of self-reported problems. These offenders did better on lower caseloads, but arguably were not very typical of offenders in general, although they broadly resembled sub-groups of offenders which had shown positive results in other comparative studies (Adams, 1961; Palmer, 1974; Shaw, 1974). The most probable explanation is that the typical content of probation casework at the time, with its focus on relationship-building and promoting insight, could be helpful to those who were distressed, anxious to change and not particularly criminal. But for probation as a general-purpose 'treat-ment' for the bulk of the criminal population, these findings were little short of disastrous, and the Home Office neither carried out nor supported much substantial research on the effectiveness of probation for the next twenty years.

Although it became, particularly in Britain, almost official doctrine that 'nothing works', some commentators regarded this as an over-reaction. In the USA Palmer (1975) argued that Martinson's verdict was too severe when a number of the studies in his review actually had positive results, and in Canada Gendreau and Ross were soon to produce their 'bibliotherapy for cynics' which brought together a number of accounts of more successful experiments (Gendreau and Ross, 1980). In Canada particularly, the 'nothing works' verdict was always contested and research on the rehabilitation of offenders continued to be pursued with both rigour and optimism. Even in Britain two Home Office studies produced results which would have been welcomed in a less sceptical era. The first was a study which pointed to characteristics of probation hostel regimes which were asso-ciated with lower rates of 'failure', i.e. early termination of placement due to absconding or a new offence (Sinclair, 1971). 'Failure' was in turn highly correlated with reconviction. The second was a study showing lower reconviction rates among released prisoners who had received thorough pre-release preparation (Shaw, 1974). Shaw's study echoed the positive findings of an earlier Scandinavian study by Berntsen and Christiansen (1965). However, instead of applying these findings to the improvement of prison welfare services, as an earlier age might have done, the Home Office sought and found a reason for not acting by carrying out another study of pre-release work with prisoners (Fowles, 1978). This was carried out in less favourable circumstances, did not have a positive result, and was widely but incorrectly presented as a 'replication' of Shaw's study. One can almost hear the sighs of official relief. Other commentators tried, largely for the benefit of practitioners, to draw more encouraging conclusions from the research (for example Raynor, 1978) but such arguments were largely ignored outside the world of social work and probation practice.

Social work and critical criminology

While traditional positivist 'scientific' criminology was shifting its ground from support for rehabilitation to demonstration that it did not occur, other kinds of criminology were emerging, which had equally alarming implications for social work and probation as traditionally understood. Positivist criminology had itself become contested ground during the 1960s: new social theorists of deviance such as Becker (1963) and Lemert (1967) had pointed out that deviancy was not a property of persons but a social status conferred and maintained by social processes, particularly by authoritative 'labelling'. As Becker put it, 'social groups create deviance by creating the rules whose infraction constitutes deviance' (Becker, 1963, p. 9). 'Diagnosis' and 'treatment' were, in this view, processes in which labels were attached and kept in place. Discussion has continued ever since about whether such theories argue that labelling is a 'cause' of crime or whether, as Becker (1973) later argued, the intention was simply to extend the focus of investigation from what was distinctive about the criminal or the deviant to what was distinctive about the social processes which led to him or her being treated as 'criminal' or 'deviant'. These discussions are largely immaterial for our current purposes: whatever its limitations as a 'theory of crime', labelling and other aspects of the 'new deviancy' theories posed a particular challenge to the helping professions in general, and to people who tried to rehabilitate offenders in particular.

This challenge arose most strongly from the concept of 'deviancy amplification' or 'secondary deviation'. Deviant social roles, it was argued, were conferred in labelling or 'signifying' processes which acted as signals both to the labelled deviant and to the rest of society that in future he or she was to be treated as a particular type of person: the conferred deviant identity became a 'master identity' which determined the nature of the social interactions in which the deviant person became involved. People diagnosed as mentally ill became inhabitants of the social world of the asylum, which confirmed their new identities (Goffman, 1961); criminals found themselves put in the company of other criminals, particularly in custodial institutions, and found obstacles such as restricted employment opportunities if they tried to return to the straight world. In other words, the actions taken in response to an initial deviation (the 'social reaction') could make it worse. Even processes and services which were intended to help could tend to confirm deviant identities and generate deviancy amplification. As Wilkins (1964) put it: 'a small initial deviation which is within the range of high probability may develop into a deviation of a very low probability' (p. 91).

Such theories were in fact derived from quite a long tradition of social enquiry going back at least to Mead's social-psychological account of the development of the self (Mead, 1934), which was not unknown to the social work tradition, even if it had been rather obscured by the vogue for psychodynamics. (For example, Mead was himself one of Jessie Taft's lecturers and PhD examiners.) However, the focus on deviant identities and the processes which generated them was largely new, and it influenced studies of a wide range of processes from juvenile court proceedings (Cicourel, 1968) to psychiatric diagnosis (Scheff, 1966; Rosenhan, 1973). In Britain, theories of deviancy amplification combined with subculture theory to produce classic studies such as Cohen's of Mods and Rockers and their seaside battles (Cohen, 1972) and Young's of soft drug users and their interaction with the police (Young, 1971).

Social scientists working in this tradition often found themselves 'on the side of' the labelled deviants in their dealings with the police and other authorities, and social workers could be described as the 'soft police' of the social control system. New deviancy theorists contributed to the new textbooks of 'radical social work' (Cohen, 1975), in which social workers could also discover that they were social tranquillizers promoting false consciousness. If they were in touch with the social movements of the New Left they could find the new deviancy there as well, as criminologists argued that the State's criminalisation of radical political movements had parallels in the criminalisation of others whose reaction to disadvantage and oppression took the form of everyday crime (Taylor and Taylor, 1968). For many social workers these ideas were a stimulus, and led both to a more politically aware approach to their work and to ideas about how social work might help to construct more positive identities for the victims of labelling (Raynor, 1985) or how it might influence the workings of systems rather than just individual people (Thorpe *et al.*, 1980). The idea of 'conscientization' was borrowed from third-world community education projects and suggested as a unifying goal for social work (Leonard, 1975), while others found a more direct social action and advocacy role in community work.

In the world of probation and rehabilitation the new theories had a particular resonance because the work concerned the (now contested) subjects of crime and criminals, but it was less clear how practice might be adapted to take account of them. With hindsight, this is not surprising: attempts to account for crime in terms of social processes only must offer at best a very partial account of behaviour that arises from a combination of individual characteristics and choices, social circumstances and social reactions. Equally the new deviancy theorists, especially when compared with the criminologists of the 1930s and 1940s, had rather few practical suggestions to offer. However, one of the implications was undoubtedly an increase in scepticism about the feasibility of rehabilitation when it was understood as a 'treatment' of some defect or pathology within the offender. The idea of deviancy as a socially attributed status ran wholly contrary to the idea of crime as a kind of disease.

Rehabilitation as injustice

Meanwhile the equation of crime with disease was also coming under attack from another direction. Wootton had argued strongly against this tendency in the late 1950s (Wootton, 1959), and moral philosophers were also arguing that conflating crime with disease involved a failure to distinguish between voluntary and involuntary actions (Flew, 1973) or a failure to distinguish between actions for which a person was morally responsible and those for which he or she could not properly be blamed (Lewis, 1971). Strawson (1968) argued that the whole vocabulary and set of ideas which we use to evaluate others' actions (including such concepts as resentment, blame, praise, gratitude, and of course guilt and innocence, punishment and reward) depend on a distinction between free actions and coerced or involuntary actions: it makes sense to regard people as responsible for the former, but not the latter. All these arguments suggested that most crimes, like most human actions, should be regarded as the result of choice or at least amenable to choice. Offenders were not the puppets of social forces or unconscious inner drives. Treating people as if they were such puppets diminished their humanity because (as Kant argued) human beings are essentially moral agents.

People working with offenders were also being led by other considerations towards the idea that most offending involves choices. In particular, social workers and probation officers in the juvenile justice field were becoming aware that for many adolescents (particularly male adolescents) some offending was to be expected as a fairly normal part of growing up; it did not mean they were seriously maladjusted, and it did not mean they were not responsible for their actions (Rutherford, 1986). They had before them an object lesson in the dangers of a 'treatment' philosophy: the 1969 Children and Young Persons Act, passed at the peak of social work's influence on British public life, deliberately blurred the difference between children who offended and children in need and aimed to look after through both a system of Supervision Orders and local authority 'Care'. Unfortunately custodial institutions for young people were not abolished as originally intended, and the first few years of the Act's operation saw a dramatic increase in their use. When researchers looked at these sentencing decisions, they found that many of the custodial sentences were suggested by social workers (Thorpe *et al.*, 1980). Concerns were also raised about children who spent long periods in residential care on 'welfare' grounds, sometimes in effect serving 'sentences' which seemed wildly disproportionate to their original offences (an example of the process described as 'net-widening' by Cohen [1985]). This attracted criticism from lawyers as well as criminologists (Morris *et al.*, 1980). (A similar campaign was waged by civil libertarians against excessive professional discretion in decisions about the mentally ill, and ultimately led to new legislation [Gostin, 1977].) Some features of the new social work and criminal justice practices which grew out of these concerns are discussed in the next chapter: what is interesting for our current purposes is that research, conceptual argument and liberal political philosophy were combining in the 1970s to undermine the rationales which had supported the rehabilitative approach in previous decades.

A particularly strong and influential set of arguments was advanced by legal scholars in America who were concerned about excessive use of indeterminate sentences, often with release dependent on professional judgments about response to 'treatment'. The American Friends' Service Committee (1971) argued that 'the individualised treatment model' was 'inconsistent with some of our most basic concepts of justice' (p. 12), and von Hirsch (1976) made a powerful case in favour of due process, determinate sentencing and a proportional relationship between the severity of a sentence and the seriousness of the offence. In particular, it was argued that people should be sentenced for what they had done, not for being a particular type of person, nor for what they might do in the future. 'Treatment', if provided at all in such a model, would be voluntary and not part of the 'sentence', although it might be under-taken during a sentence. As already noted in Chapter 2, one of the clearest short summaries of this kind of thinking was provided by Hood (1974): 'I believe that a system which arrives at the length of sentences based more on moral evaluation than on appeals to the utilitarian philosophy of deterrence and reductivism would be fairer, not neces-sarily less effective, possibly less, not more punitive and appeal to that sense of social justice on which any acceptable system of social control must be founded' (Hood, 1974, p. 7).

It is important to recognise that when people advanced this kind of argument they were not against helping offenders if they needed and requested help. They knew offenders needed more than punishment: many delinquent children are also deprived, but that does not mean delinquency and deprivation are the same. However, the fact that these arguments were mostly advanced at a time of general retrenchment in 'welfare' spending ran the risk that they would be seen as arguments for a less rehabilitative system: prisons can offer 'containment' more cheaply than rehabilitation. They were also, as later chapters will show, associated with sharp increases in custodial sentencing, particularly in America and to a lesser but still substantial extent in Britain. Nothing could have been further from the intentions of the original advocates of a movement 'back to justice'. However, it became clear that 'just deserts' had different meanings for different people. To a liberal academic it meant an end to over-intervention by professionals and an end to the passing of excessive sentences on 'treatment' grounds. To some politicians and members of the public, it seems to have meant more punishment, as they thought that the system had 'gone soft' and that people had not been receiving the level of punishment they deserved. It is ironic that probation officers and social workers were perceived as having made the system 'soft' when research actually suggested the opposite (Thorpe *et al.*, 1980): their rehabilitative efforts seem often to have had the unintended consequence of making the system more coercive and the sentences more severe. 'Treatment' is only a benefit if it is effective; failed treatment, in a criminal justice context, can put both potential victims and the offender at greater risk.

Culture, ideology and perceptions of effectiveness

As we have seen, the message that nothing worked was always contestable: the evidence was neither so clear nor so unanimous as has often been claimed. However, its dominance for a substantial period of time is a historical fact, and explanations of this dominance are likely to lie more in the territory of ideology and culture than evidence-based debate. The original proponents of 'nothing works' in the 1970s gained currency and attention for their ideas partly because they resonated with the agenda of right-wing politicians opposed to state spending on 'welfare'. The findings were used to support a neo-liberal strategy of economic individualism and 'responsibilization' (Rose, 2000), making citizens responsible for their own welfare in a free market, and relying on deterrence and punishment to maintain social discipline. Similarly, critics of social work such as Brewer and Lait (1980) used research by Joel Fischer and others to develop their idea that interfering and deluded 'progressives' were preventing the poor from sorting out their own problems. They linked this to a broader attack on the Welfare State which supported the welfare-cutting social agenda of Margaret Thatcher's new government, and like Murray's theory of the permanent underclass which perpetuates its own poverty and on which welfare spending is therefore wasted or counterproductive (Murray, 1990), they explicitly used 'nothing works' research to promote a conservative ideological position.

Garland's recent work (Garland, 2000, 2001) has tried to develop a cultural account of the shifts in attitudes to crime and punishment which accompanied and formed part of the ideological shifts of the

1970s. As well as pointing to the irony that people who campaigned for just deserts in the hope that it would lead to more lenient penal systems actually provided part of the rationale for a move to much greater severity, he draws attention to shifts in the experience of crime and in the attitudes of key middle-class groups whose support is essential for any party seeking an electoral majority. As well as increases in crime, new economic insecurities and changes in family life increased the middle classes' sense of vulnerability and risk, and these became key drivers for new and more punitive penal policies. In the past, the same social groups had shown generally liberal and progressive attitudes and had supported the developing 'penal-welfare' complex and the Welfare State; their more recent shift towards fearful insecurity and social intolerance provided the electoral base and perceived popular support for more punishment as well as more 'responsibilization'. This, of course, is no more than a brief sketch of a complex argument, but if it is correct, it helps to explain the ready acceptance since the mid-1970s of the doctrine that nothing could usefully be done to help offenders to offend less: this doctrine supported the emergent ideologies of the time. Garland's account also helps to explain why emerging evidence against the 'nothing works' position, although available, was largely ignored for so long, particularly in countries where neo-liberal right-wing governments were in control.

However, the reverse also occurred, and requires explanation. During most of the period of welfare consensus when rehabilitative ideas dominated, there was, as we have seen, little evidence available to support them, and studies going back as far as the early 1950s cast serious doubt on the effectiveness of social work and rehabilitation. The fact that this did not much dent official confidence, and that many of the public appeared to share social workers' own well-documented belief that they were doing good, also requires some explanation at the level of culture and ideology. One attempt at such an explanation was supplied by the radical critics of social work in the early 1970s (Bailey and Brake, 1975).

According to their Marxist-influenced argument, social work, whether or not it was 'effective', served an ideological function by attributing the problems of the poor to individual psychosocial malfunctioning rather than to structural inequalities in society. This concealed the true nature of 'social problems' and helped to guard against popular pressure for radical change. Similarly, social work with offenders was seen as colluding with the criminal justice system to blame individuals for social failures. This analysis can, of course, be criticised for being simplistic: not all poor people have the kind of problems which attract social work attention, not all who have such problems are poor, and offenders are not just 'victims of society' engaged in a Robin Hood strategy of primitive redistribution. They are more likely to be offending against those who are already disadvantaged. However, the idea of social work having an ideological function is useful. If 'nothing works' both contributed to and was reinforced by the political agenda of neo-liberal anti-welfarism, what in the preceding decades had been the ideological function of 'everything works'? The answer lies, we think, in the complex interplay of inequalities, Welfare State policies and professional ambition.

We have seen how early social workers in the nineteenth century were mainly engaged in a form of selective and often judgmental poor relief. Their twentieth-century successors became more interested in promoting personal change in individuals, and some of them explicitly linked this shift to their belief that the problem of poverty had largely been solved by broader social policies. For example, a widely used textbook on social services argued in 1952 that 'in the past' problems of poverty had preoccupied social reformers and social workers, whereas the most urgent social problems of the present were 'problems of psychological adjustment rather than material need' (Hall, 1952, p. 8). In America, where 'the switch-over from the economic to the psychological' (Wootton, 1959, p. 269) was even more marked than in Britain, it was argued that the increase in large-scale government-financed public relief during the depression of the 1930s almost put the voluntary casework agencies out of business, and they lost many of their public subsidies. Consequently 'they turned quite naturally to casework services that could be developed independently of relief-giving' (Miles, 1954, p. 107). In fact, the profession's assumptions about the eradication of poverty were to prove very optimistic indeed.

In Britain, the 1960s saw both the rediscovery of structural poverty persisting in spite of the Welfare State (Abel-Smith and Townsend, 1965; Coates and Silburn, 1970) and a campaign by the leaders of the social work profession to remodel the training and the organisation of social workers to comply with their 'generic' ideal. They successfully persuaded the Seebohm Committee (which neither sponsored nor carried out any empirical research) of the justice and progressive nature of their case (Seebohm, 1968) and then campaigned to persuade a sceptical government to implement the Committee's recommendations (Hall, 1976). With the exception, as we have seen, of probation, the social work services were duly unified as Social Services Departments.

This is not the place to review their subsequent history, with its mixture of significant achievements and remarkably persistent problems. What is interesting for our current purposes is that attempts to address poverty by significant redistribution were seen as too expensive, but the dramatic expansion of social work was successful in attracting significant resources. Problems of poverty were left to welfare rights workers and community workers, while social workers got on with their preferred business of casework with children and families, the mentally ill, people with disabilities and the elderly.

If Garland is right about the central importance of middle-class perceptions in influencing social and criminal justice policies, it was arguably convenient for middle-class taxpayers that at just about the same time as evidence was beginning to appear that the Welfare State had not, after all, abolished poverty, an occupational group emerged, which was eager to take on social problems by persuading individuals to change. This offered a less costly alternative to further significant redistribution of resources in favour of lower income groups. It was not particularly likely that the better-off would enquire too closely into whether the social workers' efforts were generally effective: this was, after all, still the era of faith in welfare experts. Most middle-class people wanted to see their less fortunate fellow-citizens helped, particularly if it could be done at moderate cost, but there was no reason for them to question whether it was actually happening. In the same way and for the same reasons, they might generally support the idea of treating offenders reasonably mercifully and giving them some assistance, but again the details were not their concern. Ideologies are made up of beliefs which support interests, and it was clearly in nobody's interest to disturb these arrangements by paying too much attention to evidence about the outcomes of social work. A widespread belief in the effectiveness of social work was ideologically convenient for several decades; it was not until the economic stresses of the early 1970s led to warnings about 'fiscal crisis' that middle-class people in general became more worried about their safety or their tax levels, and began to lose confidence in the State's capacity to solve problems of welfare and crime. This helped to generate support for emerging neo-liberal political movements which were more sceptical about welfare and more punitive about crime.

Finally, it would be wrong to end this chapter without some acknowledgment of rehabilitation's debt to social casework. Even Barbara Wootton (no great admirer of social work) conceded that the turn to psychology had made social workers less inquisitorial and more polite

(Wootton, 1959). But there is far more than this to write on the positive side of the balance sheet. Social casework brought to criminal justice an emphasis on getting to know offenders as individuals, listening to them, respecting their rights and opinions, and believing in their capacity for change. At its best it emphasized the importance of pro-social relationships, empathy and continuity of contact. By the mid-1970s a number of useful practical developments were also beginning to flow from research on effectiveness. One of the authors taught social workers for several years, aiming to use only practice theories which had empirical support and could be linked to practical training in demonstrably useful skills (Raynor and Vanstone, 1984). It is also well established that some helpers' methods and personal characteristics produced good outcomes regardless of their theoretical affiliation (Truax and Carkhuff, 1967). However, it would be foolish to pretend that this empirical approach was or is the dominant tendency in the profession. On the other side of the balance sheet belong theoretical credulity, over-ambition, a reluctance to give advice or direction, a preference for dealing with 'underlying' problems, a self-serving lack of interest in evidence of outcomes other than the social worker's own opinion, and perhaps most importantly an expectation that insight and understanding would on their own produce behavioural change. The empirical findings of the 1970s presented a severe challenge to these beliefs, and the next chapter explores some of the immediate consequences.

5
Adapting to the End of 'Treatment'

Rehabilitation, penal policy and the 'correctional apparatus' in the 1980s

By late 1970s, the correctional aspirations of the penal system had been seriously undermined, leaving something of a lacuna in penal policy where ideas about rehabilitation used to be. The collapse of the treatment model also created problems of legitimacy for much of the 'correctional apparatus', the supposed rehabilitative potential of which had formerly justified its existence. For those parts of the penal system which had grown up on the foundations of rehabilitative optimism, and which had been most closely associated with the delivery of rehabilitative 'treatment', the future was far from certain. In 1978 the then head of the Home Office Research Unit expressed his own concerns as follows:

> penological research carried out in the course of the last twenty years or so suggests that penal 'treatments', as we significantly describe them, do not have any reformative effect whatever other effects they may have. The dilemma is that a considerable investment has been made in various measures and services, of which the most obvious examples are custodial institutions for young adult offenders and probation and after-care services in the community for a wide variety of offenders. Are these resources simply to be abandoned on the basis of the accumulated research evidence? (Croft, 1978, p. 4)

Whilst the abandonment of the more overtly correctional or treatment-oriented parts of the system may not have been a realistic option, they nonetheless faced a crisis of identity, and the challenge to find and articulate a realistic and acceptable purpose loomed large.

The prison service

As we saw in Chapter 3, by the close of the nineteenth century prisons were increasingly being understood in terms of their reformative potential; but they were not initially conceived as instruments of rehabilitation. Thus, whilst the discrediting of 'treatment' did seem to suggest a re-think of prison's official aims, and arguably did result in a crisis of identity and morale for some prison staff (King and Elliott, 1978), it did not *by itself* result in a questioning of the legitimacy of prisons *per se*.

Nonetheless, the 1970s had been a particularly turbulent decade for the prison service, and there is no doubt that by the end of that decade it faced a crisis of legitimacy. Not only had its reformative efforts been exposed as ineffective, but a spiralling prison population had meant the housing of prisoners in increasingly squalid conditions, an increase in the number and seriousness of prison disturbances, and an escalation of industrial action (King and Morgan, 1980). This state of affairs led to the establishment, in 1978, of the May Committee. Although assembled primarily to head off the industrial relations crisis, the Committee quickly came to the conclusion that it would be sensible to go beyond its 'official' remit and address the contemporary purposes of prison.

As King and Morgan (1980) have explained, the official purposes of prison were, by this time, somewhat ambiguous. On the one hand, there was an acknowledgement that the new generation of training prisons had not brought any dramatic breakthroughs in respect of effectiveness; but on the other there seemed to be a reluctance to accept that the quest for correctional outcomes had failed. From the mid-1960s Prison Rule Number 1 had stated that 'The purpose of the training and treatment of convicted prisoners shall be to encourage and assist them to lead a good and useful life' (Home Office, 1964, quoted in King and Morgan, 1980, p. 14), but there had, by the end of the 1960s, already been an official distancing from this overarching (and ambitious) aim, in favour of an objective of 'humane containment' (Home Office, 1969). Thanks to a number of high-profile escapes, the late 1960s had also seen a re-emphasis on physical security, articulated in two official reports (Home Office, 1966; Advisory Council on the Penal System, 1968). Further, by the late 1960s, only around half of all prisoners were actually being transferred from local to training prisons in line with the 'treatment and training' philosophy, which, for King and Morgan, illustrates the ambivalence of the prison service in respect of its overarching purpose.

The May Committee formally acknowledged that in the context of imprisonment 'the rhetoric of "treatment and training" had had its day and should be replaced' (May, 1979, quoted in Cavadino and Dignan, 2002, p. 178), but the alternative of 'humane containment' was rejected as likely to 'throw out the good with the unattainable' (para. 4.24). Instead it was suggested by the May Committee that 'the purpose of the detention of convicted prisoners shall be to keep them in custody which is both secure and yet positive' (para. 4.26) – a rather bland compromise which said nothing concrete about how such a purpose might be achieved, and which was thoroughly criticised by King and Morgan. It would be another decade before the Prison Department published a revised 'Statement of Purpose', which asserted that:

> Her Majesty's Prison Service serves the public by keeping in custody those committed by the courts. Our duty is to look after them with humanity and to help them to lead law-abiding and useful lives in custody and on release. (1988: quoted in Cavadino and Dignan, 2002, p. 178)

Clearly this was a rather less ambitious purpose, allowing for the possibility of some rehabilitative impact but by no means representing a wholesale commitment to the reformative potential of prison.

The probation service and juvenile justice

As exemplar of the 'normalising sector' of the penal system (Garland, 1985), the probation service was rather more heavily implicated in the collapse of the treatment model. Having evolved on the very foundations of reformative optimism, and having been at the vanguard of developments in 'therapeutic' practice, the probation service was extremely vulnerable to calls that it could no longer command legitimacy. This was arguably already being demonstrated in that the courts' use of the probation order – the disposal most closely associated with treatment – was already in decline. Having constituted three-quarters of the service's caseload in 1961, the probation order's share had dropped to 56.5 per cent in 1971 and would fall to just 31.9 per cent by 1981 (McWilliams, 1987, p. 100). Whilst this trend appeared to be related to the availability of new measures such as the suspended sentence and community service, and the increasing popularity of the fine (Bottoms, 1977; Haxby, 1978), the demise of the probation order can also be understood as one symptom of a declining commitment to the more overtly correctional disposals.

Fortunately for the probation service, its work had already diversified significantly by the late 1970s, such that the probation order was not its sole bread and butter. One client group with which the service had become increasingly involved was offenders sentenced to custody. In the mid-1960s it had assumed responsibilities for work with prisoners both during and following custodial sentences. Following the publication of *The Organisation of After-Care* by the Advisory Council on the Treatment of Offenders (Home Office, 1963), the probation service had taken on the voluntary after-care work which had formerly been carried out by local discharged prisoners' aid societies. In addition, seconded probation officers had assumed the responsibilities of welfare officers in prisons. The Criminal Justice Act 1967 further extended the responsibilities of the service in the area of after-care by introducing parole, whilst also giving legal status to the new 'probation and after-care service'. By 1981 the proportion of after-care cases would exceed probation orders, claiming 37.6 per cent of offenders under supervision (McWilliams, 1987).

Further, the Criminal Justice Act (1972) had introduced a number of new non-custodial measures which would impact on the probation service: empowering the courts to make community service orders; introducing day training centres; and attaching supervision to certain suspended sentences. The community service order, introduced in six pilot areas in 1972 and extended nationally three years later, would steadily gain ground, accounting for 12.8 per cent of the caseload by 1981 (McWilliams, 1987).

It is hardly surprising then that, in this context, sympathetic commentators began to emphasise aspects of the service's work to which the notion of rehabilitation was not central. For example, Haxby advocated a change of title, arguing that 'community correctional service' would better convey the organisation's capacity to utilise a variety of approaches towards offenders:

> some based on social work and methods of supervision, others of a reparative type, others leaning more towards re-education, and others frankly punitive or intended to emphasise social control. (1978, p. 10)

Meanwhile, other commentators sympathetic to probation's plight were looking for ways to move practice away from the treatment model. As the decade drew to a close, Bottoms and McWilliams proposed a 'non-treatment paradigm' for probation practice, in which the various shortcomings of the treatment model were turned into an opportunity to introduce a model of practice which would be consistent with the probation service's 'traditional values of care and respect for unique individual persons, and hope for their future potential' (Bottoms and

McWilliams, 1979, p. 166). This model retained at its centre the notion of casework, but rejected the idea of the professional as 'expert'. 'The caseworker', the authors explained, 'does not begin with an assumption of client-malfunctioning; rather, he offers his unconditional help with client-defined tasks, this offer having certain definite and defined boundaries' (p. 172). Despite the authors' assertion that this formulation constituted a new paradigm for probation practice, it can be seen as an essentially adaptive adjustment of existing practices and priorities rather than a complete break with the past, namely a 're-framing' of treatment as 'help', conceived 'as something which the client rather than the caseworker defines' (p. 169), and of diagnosis as 'shared assessment'.

As well as proposing their 'alternative' model of practice, Bottoms and McWilliams's seminal article astutely identified – and capitalised upon – the Home Office's contemporary preoccupation with a growing prison population, and did not miss the opportunity to highlight the compatibility of their notion of 'help' with the 'traditional' aim of diverting appropriate offenders from custodial sentences (Home Office, 1977). Ultimately, it was this 'systemic' concern which would suggest a new direction, not just for the probation service, but for penal policy more generally. As McWilliams (1987) explained in a subsequent paper, it soon became clear that although the probation machine could not feasibly be dismantled, it *could* be directed towards another useful purpose, namely relieving the pressure on custodial institution by dealing with ever-greater numbers of offenders in the community.

Thus began what, in historical accounts of the service, has generally been characterised as an era of 'pragmatism' (McWilliams, 1987) or, more commonly, 'alternatives to custody'.

> If the emphasis of the 1970s had been on doing good without much success in demonstrating that good was being done, the 1980s were to be about avoiding harm, in particular by reducing unnecessary incarceration. This seemed a more achievable aim, and one which might commend itself on the grounds of economy even to communities or politicians who were not in sympathy with the underlying humanitarian aim. So began the era of 'alternatives to custody': probation was to be a non-custodial penalty aiming to increase its market share and reduce imprisonment, rather than a 'treatment' aiming to change people. (Raynor, 1997, p. 21)

This development in the probation context very much ran in parallel with similar changes in juvenile justice, which centred on a retreat from the 'strident interventionism' of the 1960s and 1970s (Rutherford, 1994) and the adoption of a new philosophy of 'minimal intervention' (Pitts, 1988; Bottoms *et al.*, 1990). In the last chapter we discussed the emergence, in the late 1970s, of damaging evidence of the unintended consequences of the 1969 Children & Young Persons Act, particularly the 'net widening' and 'mesh thinning' effects of Intermediate Treatment, which had been intended as a substitute for detention (Thorpe *et al.*, 1980). In the context of a growing recognition of both the potential of treatment concepts to accelerate individuals' progress up the tariff, and the inherent capacity for harm in the well-intentioned name of welfare, juvenile justice practice was by the late 1970s ripe for change. Characterised by one commentator as 'the new orthodoxy' (Jones, 1984), the approach which came to dominate juvenile justice in the 1980s essentially rejected the traditional 'welfare' approach which had tended to focus on meeting the emotional and social needs of individual young people, in favour of one which was both more concerned with responding to offending and sought to keep young offenders, as far as possible, out of court and – where prosecution was unavoidable – out of custody. This approach was encouraged by the Criminal Justice Act 1982, which placed restrictions on the custodial sentencing of young people under the age of 21.

This 'new orthodoxy' in juvenile justice clearly impacted on developments in the probation service, as the following comments of a Chief Probation Officer illustrate:

> For a long time, we felt we had a licence to do things to people for the best of all possible reasons. Partly because of the way criminal justice itself has moved and partly because of the temper of the times, we have learned that we cannot justify that. One of the first things I did in my present job was to get my committee to redefine its objectives and priorities. We started off with a set of principles, the first of which was minimum intervention. We demonstrated to the committee, regarding the juvenile justice system, that non-intervention is a very potent force for good. That colours our attitude as to who we ought to be involved with in a way that it never used to. (quoted in Rutherford, 1994, p. 21)

By the mid-1980s an official endorsement of probation's new and comparatively modest rationale had appeared, in the form of a 'Statement of National Objectives and Priorities' (SNOP) (Home Office, 1984). The national objectives and priorities laid out in SNOP confirmed the

Home Office's intent that the service's primary task should be to help reduce the prison population by providing courts with community-based alternatives. Whilst the objective of reducing the risk of reoffending did receive a mention in SNOP, the document conveyed a clear message that the effectiveness of the service would in the future be judged in terms of its ability to cope with growing numbers of offenders, rather than in terms of any 'correctional' outcomes it might be able to secure.

Back to 'justice': The rise – and compromise – of desert

In 1990 a White Paper, *Crime, Justice and Protecting the Public*, was published, setting out the Government's plans for the criminal justice system, and sentencing in particular (Home Office, 1990). These plans centred on the official replacement of rehabilitation as a principal justification for punishment, and the implementation of a version of the 'justice model' which had played a significant role in the downfall of the 'rehabilitative ideal' (American Friends Service Committee, 1971). Drawing inspiration from the recent work of academics such as von Hirsch (1976, 1986; Wasik and von Hirsch, 1988), and recalling the tradition of Beccaria's classicism, the White Paper sought to put in place a system of penalties from which sentencers could select, with the gravity of offending behaviour in mind. In place of sentences tailored to the particulars of offenders, it proposed to introduce a system-based, as far as possible, upon the notion of desert. This meant that, rather than permitting indeterminate or disproportionate sentences in the hope of reformative effects, the Government was now mindful to limit punishment to 'fit the crime'.

The White Paper proposed to continue the policy of penal reductionism, and sought to discourage the use of custody as far as possible in favour of non-custodial options such as fines, probation and community service. However, grafted onto this was a policy of 'bifurcation' (Bottoms, 1977), such that a distinction was drawn between the 'ordinary' offender, for whom a relatively lenient line was considered appropriate, and the 'really serious offender', warranting a tougher response. This meant that custodial sentences could be justified, in cases of violent and sexual offences, on the grounds of 'public protection'.

The implications of this revised penal strategy for the probation service were profound and somewhat mixed. Although continuing the policy of penal reductionism and very much promoting the use of community-based sanctions in cases where custody could be avoided, the new strategy also served to distance probation even further from the rhetoric of rehabilitation and the tradition of welfare. It did this principally by re-conceptualising community-based sanctions, such as probation orders, as part of a new 'continuum of punishment' (Mair, 1997), and making them for the first time sentences in their own right. In a framework of desert-based sentencing, the idea was that courts would opt for community sentences in cases where the offence(s) did not merit custody, but which were too serious for non-supervisory penalties such as discharges and fines. Probation, community service and super-vision orders were thus to be conceived principally as sentences capable of restricting the liberty of the offenders made subject to them, thereby representing 'just deserts' for 'middle-range' offenders. Community-based sanctions were thereby re-branded as 'punishments in the community'; tough but relatively inexpensive options for less serious offenders.

Unfortunately, however, the more 'liberal' aspirations of the 1991 Criminal Justice Act were to be short-lived. When, in 1993, the then Home Secretary Michael Howard infamously announced that 'prison works', it was clear that penal policy had moved on from the penal reductionist stance which had informed the Act. It was also abundantly clear that the effectiveness of imprisonment, for the Conservative government, had little to do with any remaining correctional aspirations, but rather its incapacitative – and possibly deterrent – effects. For a government increasingly prone to 'populist punitiveness'[1] (Bottoms, 1995; Garland, 1996), the priority of dispensing offenders' just deserts was quickly eclipsed by the rhetoric of public protection, and a priority to respond to the supposed threat posed by hordes of serious and persistent offenders. The Government's new preoccupation is neatly illustrated in that the title of the next White Paper it issued was simply entitled *Protecting the Public* (Home Office, 1996a). Intended to 'wage war on crime' (para. 13.1), this White Paper saw the increasing compromise of desert in favour of mandatory minimum (custodial) sentences for various categories of offender, including three-year imprisonment for a third conviction for domestic burglary.

In this context the probation service found itself subject to constant criticism for being too 'soft' and too closely aligned with the interests of offenders. In 1995 – the height of the period of 'populist punitiveness' – the Home Secretary took the decision to sever the link between probation and 'soft' social work, by repealing the need for probation officers to hold a social work qualification. Addressing the National Probation Conference, the Home Secretary said that he rejected 'an approach which equated punishment in the community with social work with offenders' (Travis, 1995).

The theoretical context

Despite its apparent banishment from 'official' discourse, rehabilitation was, in academic circles, subject to various attempts at 'modernisation' throughout the 1980s. However, in none of these formulations was rehabilitation understood or promoted in the guise of 'treatment', nor strictly according to a correctional model. For example, one group of commentators, dissatisfied with the justice model as a rationale of sentencing, attempted to reaffirm rehabilitation as an obligation of the state to the offender (Cullen and Gilbert, 1982; Hudson, 1987; Carlen, 1989; Rotman, 1990). Taking a slightly different tack, McWilliams and Pease advocated the resurrection of rehabilitation as the over-arching ideal directing probation practice. Noting that the probation service had for over a decade been 'conceptually spineless', they argued that the service 'stands in need of a transcendent justification for its activities. Rehabilitation may prove an attractive candidate for such a justification' (1990, p. 16). In their formulation, rehabilitation was defined as any attempt to 'help the offender return to and remain as a full member of society...a means to limit punishment to the extent pronounced' (1990, p. 15). This was a definition from which the notion of 'treatment' or 'reform' was not excluded, but to which it was not an essential component. Significantly also, rehabilitation was not advocated by McWilliams and Pease as a generic philosophy for the criminal justice system as a whole, nor as an 'alternative' to punishment.

However, these theoretical arguments made little headway, and received damaging criticism in Nellis's influential paper on probation values. Nellis (1995) argued that rehabilitation was insufficient as a value-base for probation: first, echoing Haxby almost twenty years previously, because the service had accepted responsibility for a range of activities which could not be described as *primarily* rehabilitative; secondly, because the placing of offenders' needs and interests above those of existing and potential victims 'lacks moral justification and, in the 1990s, political credibility' (1995, p. 26).

Meanwhile, another group of academics – largely but not solely outside the UK – was engaged in a more generalised debate about the fate of a correctional approach in the penal field. This debate centred on the notion of a paradigm shift and the displacement of a rehabilitative or disciplinary approach towards offenders, as conceived by Michel Foucault (1977), by a model with *risk management* at its heart. Just as Foucault had theorised a 'great transformation' from corporal punishments directed at the body and the infliction of pain, to carceral punishments directed at the correction of the mind (see Chapter 3), what was now being proposed was another transformation: this time, from rehabilitation to 'risk management'. Whilst there were distinct differences of emphasis in theoretical characterisations of risk management, this term essentially implied a shift of focus away from individuals in favour of categories or aggregates; and from a position of rehabilitative or 'transformative' optimism, in favour of more limited, 'managerial' goals. In the next section we consider the various strands of this debate before going on to evaluate their resonance in the context of British criminal justice in the 1980s and early 1990s.

From rehabilitation to risk management

In much social-theoretical work of the early 1990s, the notion of 'risk' emerged as a central cultural theme (Giddens, 1990, 1991; Beck, 1992). For theorists of the 'risk society', late modernity is characterised by dangers which are wholly man-made, which affect everyone and which cannot be converted into 'certainties', even by the expert systems on which we normally rely. Whilst considerations of risk in social theory focused in particular on low-probability, high-consequence risks such as the threat of nuclear war, the risk profile of late modern society was thought to permeate society generally, such that the identification and management of risks had become structuring principles of contemporary life. For example, Giddens noted that a significant part of 'expert' thinking was made up of *risk profiling*: that is, 'analysing what, in the current state of knowledge and in current conditions, is the distribution of risks in given milieux of action' (1991, p. 119).

In the light of the failure to eliminate crime through the application of 'correctional' interventions, the notion of 'crime as risk' seemed, to a number of academic commentators, to capture well a collective adjustment in our perception of crime, whereby offending had come to be viewed not as a sign of abnormality, but rather as a contingency or 'normal social fact' (e.g. Cohen, 1985; Reichman, 1986; Feeley and Simon, 1992; Garland, 1996). The notion of crime as risk implies a rejection, or at least a modification, of modern penality's quest to eliminate crime, in favour of more modest, 'managerial' objectives. 'A risk management approach to crime', Reichman argued, for example, 'does not offer any promises to eliminate crime by seeking out and correcting its underlying causes or by rehabilitating offenders' (1986, p. 164). Such an approach is also associated with the increasing application of actuarial or insurance-based techniques to problems of control; techniques based not on biographical knowledge of individuals, but of aggregates or populations and distributions of risk among them. A risk management approach is, thus, also associated with the de-centring of the individualised 'delinquent' subject (Foucault, 1977) in favour of 'actuarial man', who becomes the target of new preventive systems of surveillance, prediction and opportunity reduction (e.g. Cohen, 1979, 1985; Mathiesen, 1983; Shearing and Stenning, 1985; Reichman, 1986; Simon, 1987, 1988).

Whilst these correlates of a risk-based approach were initially documented in analyses of crime prevention strategies, similar developments began, from the mid-1980s, to be observed in respect of the treatment of offenders in the penal realm (e.g. Cohen, 1985; Simon, 1987, 1988, 1993; Feeley and Simon, 1992, 1994). One of the first to bring the penal sphere under the umbrella of such analyses was Cohen, who observed, across the whole spectrum of crime control strategies, a 'new behaviourism', characterised by 'an uneven move away from internal states to external behaviour, from causes to consequences, from individuals to categories or environments' (1985, p. 154). As part of this shift Cohen observed a retreat from Freudian-derived 'inner states' models of offender rehabilitation, in favour of a less ambitious vision of behaviour modification, content to 'settle for sullen citizens, performing their duties, functioning with social skills, and not having any insights' (1985, p. 151).

But whilst Cohen posited a partial retreat from correctional interventions, American criminologists Feeley and Simon (1992, 1994) took the argument a stage further. Their 'new penology' thesis rested on a wholesale shift from a penality characterised as clinical, individualised and treatment-oriented, in favour of a 'new' penality characterised, in contrast, as actuarial, managerial and control-oriented. In summary, the new penology, according to Feeley and Simon, 'is managerial, not transformative... It seeks to *regulate* levels of deviance, not intervene or respond to individual deviants' (1992, p. 452). Feeley and Simon's new penology thesis went further than Cohen's 'new behaviourism' in two important respects. First, it argued that the individual, *even as a locus of behaviour*, was becoming increasingly invisible as the 'penal enterprise' shifted away from a concern with reforming the individual to managing quantities of offenders. Secondly, it predicted not just a lowering of expectations about the potential of criminal sanctions, as Cohen's analysis implied; but rather the wholesale rejection of any overarching or 'superordinate' goals by the penal apparatus – rehabilitation included.

In the context of an actuarial or risk management regime, Feeley and Simon argued, new justifications for existing practices had emerged. No longer were imprisonment or supervisory penalties anchored in aspirations to correct or rehabilitate offenders: the prison had become little more than a warehouse for the highest risk groups, whilst community-based sanctions simply offered opportunities to maintain control over lower risk offenders for whom custody was judged unnecessary or too expensive. Whether in the context of custody or community-based supervision, they argued that the technology of risk assessment would increasingly be relied upon to perform the task of classification; serving simply to 'sort' individuals into groups according to 'the degree of control warranted by their risk profiles' (1992, p. 459).

A paradigm shift?

It is relatively easy to identify many of the trends identified under the umbrella of a 'risk management' approach towards offenders in the UK during the period in question. For example, there is little doubt that, during the 1980s and the early part of the 1990s, the ambitions of penal policy took a nosedive. This general trend is succinctly described by Garland (1996). We have already documented how the official goals of the prison and probation services underwent change in this period, essentially distancing themselves from their former rehabilitative ambitions. As Garland (1996) has pointed out, the White Paper and the Criminal Justice Act which followed it in 1991 consolidated a lowering of penal expectations which had been in train since the late 1970s, ushering in the apparently achievable goal of punishment in line with desert in place of much more ambitious 'correctional' goals. Indeed, the White Paper did contain numerous references to such a recalibration of penal objectives, noting, for example, that the aim of its proposals was, quite simply, 'better justice ... so that convicted criminals get their "just desserts" ' [sic] (Home Office, 1990, para. 1.6) and, elsewhere, warning against too much investment in criminal sanctions as a means to reduce crime (Home Office, 1990, para. 1.8). Certainly, then, new justifications for existing penal institutions and practices emerged, and these tended to be rather less ambitious than in previous decades.

This period also witnessed the emergence of new ways of thinking about and responding to offenders, which seemed to resonate with much of the theoretical work discussed above. First of all, in the probation context, several commentators noted that a discourse based on the 'management' – rather than the 'treatment' – of offenders had begun to emerge by the late 1980s. For example, Home Office proposals to enable the probation service to work in partnership with non-statutory agencies in the provision of services for offenders (Home Office, 1991) were singled out as heralding a new management model of practice. In this context Nellis (1990) predicted that

The task of the social worker/probation officer will be to arrange packages of care (or is it punishment in the case of probation officers?) by drawing on the resources of various agencies in the community . . . [bringing about] a radically different pic n' mix style of managing offenders. (1990, p. 42)

Such a model of practice, with the probation officer acting as a 'broker' as well as a provider of services, potentially presented a significant challenge to the tradition of practice based on the establishment of a 'therapeutic' relationship between the individual offender and his or her officer. This implied shift from 'casework' to 'case management', though virtually undocumented in the literature, was set to become an accepted part of probation discourse; as indeed was the notion of 'risk' as a principle for organising and responding to offenders (Kemshall, 1998).

The notion of risk and, indeed, the practice of risk assessment were, in the 1990s, not entirely new to the probation service. For example, in the so-called era of 'alternatives to custody', the service had been much interested in *risk of custody*: that is, the identification of those offenders for whom a custodial sentence was a likely option and who could, therefore, be legitimately targeted for a community-based 'alternative' (Bale, 1987). However, the legislative changes of the early 1990s brought about a fundamental re-conceptualisation of risk *as an attribute of offenders*. As we explained earlier in this chapter, the 1991 Criminal Justice Act introduced the notion of 'public protection' as a justification for the passing of custodial sentences in cases of violent and sexual offences. This served to focus attention on offenders' potential to cause 'serious harm', and placed a new onus on the probation service to provide, in appropriate cases, evidence of risk to the public. Indeed, National Standards for the supervision of offenders in the community stated that 'the management of risk, both of serious re-offending and of serious harm to the public, where relevant, is an important part of the work of the service' (Home Office, 1992, p. 32).

A position statement from the Association of Chief Officers of Probation (ACOP, 1994), issued in the same year as the report of the inquiry into the care of Christopher Clunis[2] (Richie *et al.*, 1994), arguably marked the beginning of an explicit focus on risk in the probation service. This statement recognised a 'need for a culture shift within services to a more active recognition of responsibility for assessing the risk to the public and its subsequent management' (ACOP, 1994, p. 1). A subsequent report of a thematic inspection of the service's work with 'dangerous' offenders (HMIP, 1995) conveyed a clear message that the supervision of potentially dangerous offenders should take priority over that of 'other' offenders. It also argued that joint policies, procedures and strategies should be agreed with other agencies with responsibility for public protection: particularly the police, but also social services, health providers, special hospitals, prisons and voluntary agencies. At the heart of this multi-agency model was the notion of *intelligence*, or the sharing of information, with the Inspectorate report recommending, for example, that there should be open exchange of information between agencies about individual cases (HMIP, 1995, 1998).

The lack of predictive precision in relation to 'dangerousness', coupled with the need to present a credible image as an organisation capable of protecting the public, led to the establishment of a central reporting system requiring probation services to notify the Home Office in the event of an offender under supervision being charged with a serious offence (Home Office, 1995a). This system was to serve as a defence against criticism in relation to the handling of such cases, in that it allowed the service to demonstrate either that such an incident was unpredictable, or that an offender assessed as potentially dangerous was supervised to a high standard and to a plan based on an understanding of known risk factors. The importance of this notion of the 'defensible position' (Kemshall, 1998, pp. 68–69) was reiterated by the Inspectorate of probation, as follows:

> Working with potentially dangerous offenders is almost always about prevention and damage limitation rather than cure. Even the best practice cannot guarantee a person will not seriously harm someone else. Staff must do the best they can and if this fails to prevent an offender seriously harming another person they should not be blamed... The aim is to reach a 'defensible position' – one which would stand up to scrutiny if the handling of the case were to be investigated. (HMIP, 1997, pp. 231, 245)

Clearly, then, this aspect of the service's work was not conceived or understood as being about 'treatment' or 'correction'; but rather about the management and containment of risk.

Meanwhile, the publication of revised national standards had generalised the concept of public protection beyond 'dangerousness' to the broader category of 're-offending' (Home Office, 1995b, p. 2). For probation officers this meant a new requirement to include an explicit risk assessment in all its pre-sentence reports (PSRs), not just in cases of violent or sexual offences. The new national standards also emphasised risk assessment as a regular requirement in relation to all offenders under supervision. As an aid to risk assessments, both in PSRs and 'more generally', the Home Office introduced an actuarial risk assessment tool, the Offender Group Reconviction Scale (OGRS), to probation services in 1996 (Home Office, 1996b). Developed on the basis of centrally held information about the demographic characteristics and offending histories of a large sample of offenders, OGRS provided an estimate of the statistical likelihood of reconviction in a two-year period. OGRS thereby seemed to confirm the emergence of an actuarial approach towards offender management, as predicted by the risk management theorists.

As Kemshall rightly observed, by 1997 the profile of risk in the probation service was such that risk had become 'the key classificatory mechanism for organising all probation work' (1998, p. 80). In a similar vein, Kemshall *et al.* (1997) argued convincingly that the policies and practices of the probation service – in common with the personal social services – were increasingly concerned with the issue of risk, and that the assessment, management and monitoring of risks were rapidly becoming their dominant *raison d'être*. They went on to argue that the centrality of risk in the probation service was evidence that it was part of the 'new penology' described by Feeley and Simon,

> engaged in a transition from the traditional concerns of the rehabilitative and welfare arm of criminal justice [. . .] and the language of 'need' to an agency of crime control concerned with the accurate prediction and effective management of offender risk. (Kemshall *et al.*, 1997, p. 217)

But to what extent was this the case?

Whither rehabilitation?

In the preceding section we outlined a number of developments which seemed to suggest the emergence of a risk management approach, particularly in the probation context. But we need to exercise caution here, since it is one thing to identify signs of such an approach, and quite another to argue that such developments actually rendered rehabilitation redundant, albeit that some theorists suggest this is so. What, then, was the fate of rehabilitation during the 1980s and early 1990s?

Thanks to a dearth of research on the views and working practices of penal practitioners during this period, it is actually extremely difficult to chart the fate of the 'rehabilitative ideal' in the realm of practice. In attempting to reconstruct developments in penal practice 'after treatment', it is therefore tempting to take the key messages from theory and policy at face value and simply 'read off' developments in practice. However, we should be wary of such an approach. Indeed, several commentators have noted that it is not at all clear how far rehabilitation as an objective was actually abandoned by practitioners – most notably probation officers – following its banishment from 'official' discourse. For example, writing in 1980, one commentator posed the question 'is the rehabilitative ideal really dead?':

> The rehabilitative ideal seems to me to be a remarkably lively corpse... even the most cursory examination of the recommendations of probation made in social enquiry reports, or a reading of the journal of the National Association of Probation Officers, *Probation Journal*, or a conversation with the most experienced probation officers makes it abundantly clear that the rehabilitative ideal is not a lost cause in the probation service. (Pease, 1980, p. 149)

Pease is not alone in proposing that 'nothing works' research actually made little difference to practitioners. Vanstone (2000), for example, has argued that negative research findings remained largely invisible to practitioners, and that in the process of implementing the 'post-treatment' policy of diversion from custody, they continued to work towards 'rehabilitative outcomes' when community supervision was awarded by the courts (see also Willis, 1983).

But what about the style, or methods, of rehabilitation? Did a continued concern with rehabilitation mean a continuation of the practices associated with the 'old' treatment model; or did 'rehabilitative practice' undergo a transformation? In this context it is worth recalling Cohen's thesis, concerning the rise of a 'new behaviourism' – a style of rehabilitation less concerned with changing the offender's personality or 'psyche' than with attempting to change his or her anti-social or 'delinquent' behaviour. This argument does appear to have some credence. For example, we have already noted in this chapter that social work with juvenile offenders was, by the mid-1980s, already becoming more offence-focused, and the 'offending behaviour' approach suggested by McGuire and Priestley (1985) appears to have reinforced this trend in practice with adult offenders (Vanstone, 1993, 2000) (see also Chapter 6). So too, arguably, did a 1988 Home Office circular, *Tackling Offending: An Action Plan*, which explicitly acknowledged the successes of offence-focused programmes of supervision for juvenile offenders and sought to encourage the use of intensive probation programmes run along similar lines for young adult offenders.[3]

However, Vanstone's recent analysis of probation's practice history cautions against an over-simplified account of developments. On the basis of interviews with long-serving practitioners and documentary analysis, Vanstone (2004) has argued convincingly that the break with the treatment model was not a clean one, and practice did not change overnight. For example, although there was an increasing tendency to understand the offender as an active 'partner' in the supervision process, probation officers continued to rely largely on psychological models and to focus on (changing) individuals rather than their environments. Further, Vanstone argues that behaviourism was only one of a number of theories which came to influence practice during the 1980s: other influential models included the task-centred casework approach, neuro-linguistic programming, transactional analysis and the 'non-treatment paradigm', discussed earlier. It seems, then, fair to say that no one model dominated during the 1980s, and practitioners continued to enjoy a high degree of discretion in respect of the style(s) of practice they chose.

It seems clear, then, that in the realm of penal practice, rehabilitative impulses did not evaporate in wake of the treatment model, albeit that practice itself did change, gradually and unevenly. To what extent, then, did practice conflict with policy during the period in question? Certainly the 'new penology' thesis suggests that it did. However, not all theorists have declared rehabilitative impulses to be dead in the context of penal policy 'after treatment'. In two early pieces, Bottoms (1977, 1983) predicted that far from dying out altogether, rehabilitation was likely to retain *partial* (as opposed to wholesale) legitimacy: that is, in relation to selective groups of offenders. Commenting initially on contemporary proposals for the preventive detention of 'dangerous' offenders, Bottoms (1977) anticipated a tendency towards 'bifurcation' in penal policy which, he suggested, would be largely motivated by the economic imperative to reduce the prison population. In a subsequent paper, he observed further signs of a more differentiated approach towards offenders, in the guise of a rapid growth of 'non-disciplinary' penalties, namely the fine, compensation orders and community service. On the basis of this observed trend, Bottoms advanced a thesis based on a *differentiation of discipline*: that is, the drawing of a distinction between cases deemed to require disciplinary (or correctional) punishments and those not warranting such interventions.

This notion of a differentiation of discipline, though hardly discussed in the literature, had much to recommend it, and was arguably borne out by subsequent developments. For example, we have already discussed how, in the juvenile justice context, Intermediate Treatment was not abandoned, but rather came to be more carefully targeted as a 'heavy-end' disposal for young offenders at risk of custody (Bottoms *et al.*, 1990). More generally, a careful reading of policy documents published in the period under consideration reveals that whilst there was a definite distancing from rehabilitation as an 'overarching' objective, this did not constitute a *wholesale* rejection of correctional aspirations for the penal system. The 1990 White Paper provides a good example of this tendency. On the one hand, the White Paper does rather baldly reject the rehabilitative potential of prison, arguing that 'nobody now regards imprisonment, in itself, as an effective means of reform for most prisoners' (Home Office, 1990, para. 2.7). However, the same degree of pessimism is not extended to community-based penalties: sentences like probation orders, it is argued, still hold out the best prospects for 'reforming' offenders.[4] (Home Office, 1990, para. 2.7). In this context the White Paper makes it clear that despite the collapse of rehabilitation as a penal 'meta-narrative', or overriding purpose, the rehabilitation of offenders is still considered a desirable objective, with the important proviso 'if it can be achieved' (Home Office, 1990, para. 2.6).

It can be argued, then, that in the 'post-treatment' era, penal policy – far from barring efforts to rehabilitate offenders – offered an essentially *permissive context* within which practitioners were able to experiment with new methods and continue to strive towards rehabilitative outcomes with at least some offenders. For example, the 1980s drive to render community-based sentences credible and demanding alternatives to prison arguably created a space in which practitioners were able to develop high-tariff, 'rehabilitative' programmes for offenders at risk of custody (Raynor, 1995). Subsequent preoccupations with risk and public protection, far from militating against rehabilitation, actually served to encourage practitioners' efforts to bring about correctional outcomes, albeit only in respect of those offenders deemed worthy of the investment (Robinson, 1999). Writing in the mid-1990s, Garland observed that:

> Rehabilitation no longer claims to be the overriding purpose of the [penal] system, or even of traditionally rehabilitative interventions such as probation...It is no longer viewed as a general all-purpose prescription, but is instead targeted upon those individuals and

groups most likely to make cost-effective use of this expensive
service...It has become a (contingent) means, not an end in itself,
and if it doesn't work, one can immediately revert to other, more
effective means, such as custody and incapacitation. (Garland,
1997a, p. 6)

Garland's summary of the status of rehabilitation in the mid-1990s
clearly owes much to Bottoms's 'dispersal of discipline' thesis, capturing
neatly the altered status of rehabilitation as a penal objective. For Garland,
rehabilitation has metamorphosed into one of a number of possible
strategies centred on the enhancement of public protection.

Conclusion

Assessing the salience of the 'new penology' thesis outside the US, a
number of commentators have argued that Feeley and Simon not only
overstate the dominance of actuarial power but also ignore continued
rehabilitative (and punitive) impulses in crime control (O'Malley, 1992;
Bottoms, 1995; Garland, 1995, 1997b). These commentators have each
argued cogently that apparently contradictory penal strategies can and
do coexist, such that the increased salience of one such strategy (for
example, actuarialism or punitiveness) at a particular point in time does
not negate alternative strategies, which may continue to operate in
tandem. It is our contention that, far from dying out during the 1980s
and early 90s, rehabilitation adapted and survived. We have argued that
both the strategy of 'penal reductionism' and the subsequent quest to
make community penalties tougher and more demanding contributed
to the development of rehabilitative interventions in both juvenile
justice and probation contexts. In the following chapter we will see
how this provided an important springboard for a more concerted
effort to reinstate rehabilitative optimism in the penal system, and
re-emphasise a correctional model of rehabilitation.

We should not, however, conclude this chapter without reference to
some of the broader notions of rehabilitation described in Chapter 1.
Clearly, the extent to which rehabilitation is judged to be an element of
penal practice depends upon the model of rehabilitation in question.
Whilst we have, in this chapter, arguably been preoccupied with changing visions of 'correctionalism', we should not neglect the parallel
emergence of new roles and sanctions, a number of which fit the broad
definition of rehabilitation discussed in Chapter 1. For example, we
discussed, at the beginning of this chapter, the new responsibilities

adopted by the probation service during the 1960s and 1970s, and this diversification brings into question the easy assertion that probation was distancing itself from rehabilitation on a wholesale basis. Reflecting on the models of rehabilitation outlined in Chapter 1, it seems clear that whilst probation may well have been partially retreating from an overtly 'correctional' approach, it was at the same time embracing a more 'reintegrative' model, assisting increasing numbers of offenders in the transition from custody to the community following release from prison. It is also well worth noting that community service orders were conceived as sanctions capable of being 'rehabilitative' (as well as retributive, educative, reparative and less expensive than prison) (Advisory Council on the Penal System, 1970), arguably drawing inspiration from a long history of efforts to bring about positive change by inculcating the disciplinary 'habits of industry' (see Chapter 3).

6

The New Rehabilitation: 'What Works' and Corrections at the End of the Twentieth Century

The previous chapter has shown how practitioners and correctional services adapted to the doctrine that 'nothing works' – sometimes by developing new roles or new understandings of their task; sometimes by trying to develop more effective forms of practice; sometimes by looking for beneficial outcomes through changes in sentencing patterns rather than through changes in offenders' behaviour; and mostly, perhaps, by not regarding the 'nothing works' research as the conclusive verdict on their efforts. As we saw in Chapter 4, a tendency towards lack of interest in research on outcomes was one of the less helpful traditions of the social work profession, but it did provide a way for practitioners to keep going when the research seemed to be consistently against them. The price, perhaps, was a greater difficulty in promoting an interest in research and in research-led practice when more helpful research began to be available.

The main purpose of this chapter is to summarise what happened to correctional services, and to probation services in particular, when the consensus of research began to change and services tried to apply new research-based models of how to reduce offenders' tendency to re-offend. Particular attention is paid to the probation service in England and Wales, which has been the location of the most comprehensive attempt anywhere in the world to create a step-change in the effectiveness of rehabilitative services by implementing the lessons of the new research. Such a massive commitment to evidence-based practice and the accompanying organisational changes was watched carefully by correctional services in many countries: as one member of the Conference Permanente Europeenne de la Probation put it:

> The Probation service in England and Wales has always been in the vanguard in these developments, and many other European studies are watching it like a hawk, ready to accept that which seems to be working and to criticise that which isn't. (Ploeg, 2003, p. 8)

This chapter describes the nature of the evidence and where it came from; the attempt to implement 'what works' in Britain; the results achieved, so far as these can yet be ascertained; the reasons why initial high expectations were not met, and finally, whether on balance the outcomes of the experiment have been worth the risk. In the course of exploring this history, we also consider whether evidence has been correctly understood or correctly used; whether the timescale of the experiment created impossible demands, and how far the problems of implementation were increased by the highly politicised context in which criminal justice services now have to operate.

The lessons of the 1980s

The previous chapter showed how 'nothing works' became a dominant belief in Britain. Elsewhere, and particularly across the Atlantic, some similar processes were at work, but the more pluralistic research environment allowed the continuation of research which ran counter to the received orthodoxies of the time. The 1975 research review by Lipton, Martinson and Wilks (which prompted Martinson's headline-grabbing 'nothing works' article of 1974) did not in fact reach uniformly negative conclusions about the studies it examined. For example, both casework and counselling were found to have produced some positive results with offenders in some circumstances, even though their overall impact was small:

> to the degree that casework and individual counselling provided to offenders in the community is directed towards their immediate problems, it may be associated with reduction in recidivism rates. Unless this counselling leads to solution of problems such as housing, finances, jobs or illness which have high priority for offenders, it is unlikely to have any impact upon future criminal behaviour. (Lipton, Martinson and Wilks, 1975, p. 572)

Later re-examinations of the same studies (for example by Thornton, 1987) found a number of positive results. Other reviews also began to point to more encouraging conclusions: for example, Blackburn (1980) reviewed a more recent group of studies than those covered by Lipton, Martinson and Wilks, and found that while few studies met rigorous methodological standards, those which did meet them showed reductions in recidivism. In Canada, Gendreau and Ross identified a number of studies with positive outcomes and reviewed them as 'bibliotherapy for cynics' (Gendreau and Ross, 1980), an early example of the many positive contributions Canadian researchers were to make to the literature of effective rehabilitation. Perhaps most surprisingly, Martinson himself published a reappraisal of his earlier conclusion that 'nothing works' (Martinson, 1979), arguing that this view was simply incorrect.

The debate which seemed largely closed, at least in official circles, in Britain continued in other countries, with positive findings emerging from Europe (for an early example, see Berntsen and Christiansen, 1965) as well as from North America. Even in Britain, some studies showed positive results: for example, enhanced input from prison welfare officers prior to release led, in a well-designed study, to lower reconvictions than in a randomly allocated control group (Shaw, 1974) and probation hostels with firm but caring wardens showed less 'failure'[1] among their residents than other hostels (Sinclair, 1971). Shaw's study in particular prefigured current concerns about the 'resettlement' of prisoners. However, these results were seen as anomalous and did little, in Britain, to dent the 'nothing works' consensus, particularly as the Home Office carried out and published a random allocation study comparing the reconviction rates of probationers supervised on low caseloads with those supervised on normal caseloads. This study, discussed in more detail in Chapter 4, was known as IMPACT (Intensive Matched Probation and After-Care Treatment – Folkard *et al.*, 1976) and was planned as the definitive test of probation officers' treatment claims. It showed that offenders supervised more intensively reconvicted slightly but not significantly more within twelve months than those supervised on normal caseloads.[2] After IMPACT the Home Office carried out virtually no further research until the late 1990s concerning the impact of probation services on reconviction: such research as was done concerned system impacts such as diversion from custody, for the reasons described in the previous chapter.

However, probation services still needed to think about effectiveness. Community service seemed to be a marketing success, but the market share of probation orders was falling through most of the 1970s. Probation orders which could be seen as a credible disposal for more serious offenders would need to offer more demanding and, if possible, effective programmes of supervision. The 1982 Criminal Justice Act encouraged the inclusion of additional requirements in probation orders to facilitate special programmes, but new larger packages needed more content. Juvenile justice specialists were already developing intensive Intermediate Treatment (that is, programmes of supervised activity undertaken as part of a supervision order) with often quite sophisticated programme content (e.g. Denman, 1982), and probation services began to follow suit with various forms of enhanced probation, despite the reservations of some of their staff (Drakeford, 1983). Occasionally these involved an almost bizarre degree of emphasis on control and discipline (Kent Probation and After-Care Service, 1981) but more often they looked for content which seemed likely to be useful to offenders and was intended to reduce their offending. Among these piecemeal and usually unevaluated developments, a few projects took research seriously enough to involve local academics in what became a new style of evaluative study.

Such local projects were typically concerned both with 'market shares' and with impacts on offending, and the combination of modest scale and locally based research allowed for adequate follow-up of both sentencing trends and the behaviour of offenders. Two studies in particular, carried out during the 1980s and published towards the end of the decade (Raynor, 1988; Roberts, 1989) were able to address some of the issues about effectiveness which had almost vanished from the national research agenda, and in both cases some diversion from custody and some impact on reconviction could be reasonably convincingly demonstrated. One of the studies (Raynor, 1988) was also able to document changes in offenders' self-reported problems, and suggested a link between reductions in these and subsequent reductions in offending. As these and other researchers increasingly questioned the assumption that 'nothing works' the Government began to outline an enhanced 'centre stage' role for the probation service in the new policy proposals which were leading towards the 1991 Criminal Justice Act, and the prospects for a more effective and rehabilitative approach to probation began to look a little brighter.

A return to optimism about rehabilitation was beginning to be evident in the prison service as well as in the probation service: for example, people became aware of Canadian experiments in running prison-based cognitive skills programmes. Around this time probation services began to be discouraged by Home Office officials from using the language of 'alternatives to custody': community sentences and prisons were no longer to be in competition, but were targeting different levels of seriousness. Also the two services were meant to be cooperating in new forms of throughcare for prisoners (Maguire and Raynor, 1997): this would be difficult if the probation service continued to see its distinctive mission as to save people from the attentions of the prison service. In practice, the 1991 Act turned out to be genuinely decarcerative in the short term, securing unprecedented reductions in the use of custodial sentences during the few months of 1992 and 1993 in which it was allowed to operate as intended, before politicians lost their nerve and repealed key sections. However, the decisive shift away from the language of 'alternatives to custody' turned out to be one of the more enduring legacies of the 1991 Act, to the extent that much of what was learned from successful diversionary research and practice in the 1980s was all but forgotten.

Concurrently with the theoretical debates reviewed in the previous chapter, a number of practical achievements deserve attention, and there are useful lessons to be learned from them. The first lesson is that diversion was not abandoned because it failed: on the contrary, among juveniles in particular it achieved substantial reductions in custodial sentencing and the almost complete disappearance of residential care for juvenile offenders during the 1980s (Smith, 1995). Although concentrating too exclusively on diversion may involve doing too little about some persistent offenders, in its own terms the policy of 'alternatives to custody' for juvenile offenders succeeded until public and political opinion in the 1990s began to favour a more punitive approach and the numbers of juvenile offenders in custody began to rise again. Among adults, community penalties in the 1980s successfully moved 'up-tariff' to accommodate a riskier and more heavily convicted group of offenders: by 1989 only 14 per cent of community service orders were being made on first offenders, but by 1999, with diversion from custody no longer a policy priority and with probation services financed according to the number of people they supervised, the figure had drifted back to 42 per cent. Probation and 'combination' orders[3] showed similar trends (Raynor, 1998a), in spite of evidence that for first offenders a fine is much less likely to be followed by reconviction than a probation order (Walker *et al.*, 1981). This recent drift towards a less criminal community sentence population has coincided with growth in imprisonment and reduction in the use of fines.

Some commentators have argued that the creation of 'alternatives to custody' *necessarily* has the unintended effect of increased recruitment to the custodial part of the system. Cohen (1985) in particular has argued that the creation of less severe sentencing options often serves simply to draw more people into the net of social control measures ('net-widening') and that this exposes them to more severe sanctions when lower-tariff measures 'fail' (a process known to youth justice staff as 'tariff escalation'). Worries of this kind accompanied, for example, the introduction of specific requirements in probation orders intended to operate as 'alternatives to custody' (Drakeford, 1983). Were probationers being 'set up to fail' by over-demanding requirements which would lead to custodial sentences on breach? Concerns about net-widening were far from unfounded: for example, juvenile justice researchers in the late 1970s found that virtually all custodial sentences passed on juvenile offenders were recommended by social workers (Thorpe *et al.*, 1980) and that supervision orders on first offenders could, if breached, lead to a custodial sentence much earlier in the offending career than would otherwise have been expected. Much was also made of the finding that suspended prison sentences appeared to have led to an increase in the prison population rather than a reduction (Bottoms, 1980b). The reason seemed to be that sentencers were passing suspended prison sentences in cases where they would not otherwise have sentenced custodially, and when some offenders re-offended the resulting custodial sentence was lengthened by the addition of the suspended term, leading to a substantially longer first custodial sentence than they might otherwise have expected.

However, there was not much other evidence of a general tendency towards tariff escalation.[4] Studies of some successful 'alternative' projects emphasized appropriate targeting to ensure that only those genuinely at risk of custodial sentences became involved, and appropriate enforcement strategies, agreed with the courts, aimed to ensure that the outcome of enforcement action would be a return to the project in as many cases as possible. For example, seven out of ten breach cases in one successful project (Raynor, 1988) resulted in returns to the project rather than custodial sentences, despite the 'high-tariff' nature of the offenders concerned. In short, the unintended outcomes identified by Cohen can happen, but they can also be avoided by conscious attention to targeting and to appropriate proposals in pre-sentence reports.

Given such safeguards the existence of additional sentencing options can be helpful, and appropriate targeting has sometimes proved quite feasible: for instance, it appeared to be achieved in several early 'intensive probation' experiments (Mair *et al.*, 1994) which recruited offenders clearly at risk of custodial sentences. Early studies of community service also found that 45–50 per cent of such orders appeared to be made instead of prison (Pease *et al.*, 1977). Even in a generally punitive climate local criminal justice systems have some relative autonomy and good results are possible. Unfortunately the determinism of Cohen's account may have contributed to the resistance shown by some probation officers to the inclusion and enforcement of additional requirements in probation orders. However, the evidence of the 1980s indicates that given good information and a systematic approach, community sentences can successfully be used on a considerable scale where otherwise custodial sentences would be passed.

The practices which enabled this to happen were in some ways precursors of the 'what works' movement, since they were evidence-based and encouraged an empirical approach to the outcomes of practice; however, their focus on outcomes in the criminal justice system ceased to be of primary professional interest for two reasons. The first was that the rhetoric of 'alternatives to custody' was firmly discouraged by a combination of 'prison works' pronouncements (Howard, 1993) and the emerging need to work more closely with prisons. The second was the re-emergence of interest in methods which showed some promise of actually influencing the behaviour and future offending of people under supervision. This restoration of a form of the traditional rehabilitative focus tended to replace awareness of system impacts, and although some commentators have argued that probation services can have a dual focus, aiming to influence both the behaviour of offenders and the operation of criminal justice systems (Raynor *et al.*, 1994), it seems in practice that many of the lessons of the 1980s have been forgotten in the process of learning the lessons of the 1990s. Some unfortunate results of this forgetfulness are discussed later in this chapter.

The new evidence and its origins

During the ascendancy of the 'nothing works' doctrine, practitioners had to find their own sources of optimism and belief in what they were doing. This actually led to a period of creativity and enthusiasm in the development of new methods and approaches. These often depended on the enthusiasm of a few probation officers, but within this pluralistic approach to practice can be found the fore-runners of modern evidence-based approaches (Vanstone, 2000). For example, the first attempts in Britain to introduce principles based on learning theory into mainstream probation practice came about through the work of two psychologists, James McGuire and Philip Priestley, who were to play a major role in the development of evidence-based practice. During the mid-1970s they offered courses to probation officers on methods they published in a handbook of 'social skills and personal problem solving' (Priestley *et al.*, 1978), and were also involved in the implementation of these methods in probation day centres and prisons. The methods used were based on 'life skills' and problem-solving techniques which were increasingly being taught by clinical psychologists to their patients or clients, and many of them were already in use in a Canadian project for unemployed young people called Saskatchewan Newstart.

Later, when evaluation of the work in prisons indicated little effect on subsequent offending except among violent offenders, Priestley and McGuire developed and taught an approach based more specifically on the analysis and modification of offending behaviour (Priestley *et al.*, 1984; McGuire and Priestley, 1985). As we have seen, a brief official flirtation with emerging ideas about effective rehabilitation occurred during the preparation of the 1991 Criminal Justice Act, but strong government endorsement of an evidence-based rehabilitative approach had to await the election of a new Labour government in 1997. By then, significant strands of evidence had emerged in Britain and elsewhere to support the methods and techniques of a new model of rehabilitation. This evidence came mainly from three strands of research: the development of psychological approaches to offending based on social learning theory; the appearance of systematic reviews of research on methods of rehabilitative intervention with offenders, and local evaluations of some structured programmes which appeared to demonstrate feasibility. Whilst there is no substitute for reading the original studies, the main features of each strand are summarised below.

The psychology of offending: Learning to be anti-social

The first strand of influential research comprises the work of psychological criminologists, many of them Canadian, who have emphasized the role of social learning and of thinking or cognition in the development and maintenance of offending. A feature of the environment in which this work has developed is that a number of key individuals have combined significant academic or research contributions with experience as practitioners within the criminal justice system, in a way which has always proved difficult in Britain (examples are Don Andrews, Jim Bonta, Paul Gendreau, Frank Porporino, Liz Fabiano, Robert Ross). A clear statement of a social learning approach to offending is provided by, among others, Andrews and Bonta's textbook *The Psychology of Criminal Conduct* (1998). This sets out an integrated theory of offending which connects social disadvantage, personality traits, thinking styles and social strategies into a model of how offending occurs and continues. For example, adverse social factors such as poverty and lack of opportunities can make it difficult for parents to provide a consistent and supportive environment for children. Personality characteristics such as impulsiveness or risk-seeking, perhaps reinforced by peer expectations, are likely to limit the benefits gained from formal education, while exposure to illegitimate opportunities and positive peer support for delinquency will make offending an attractive option. Add to this poor social skills and problem-solving abilities, perhaps due to rigid thinking, lack of awareness of alternatives and difficulty in appreciating or taking into account the views and needs of others, and offending becomes likely; add a strong possibility of getting away with the offence, or alternatively a penal response which stigmatises and excludes without addressing any of these problems, and continued offending becomes still more likely. Some of these 'risk factors' are 'static', meaning they have already happened and cannot now be changed (for example, a pattern of offending in the past) while others are, at least in principle, 'dynamic' or potentially subject to change (such as current attitudes, beliefs, behaviour and opportunities).

This kind of model also suggests a process of intervention based on trying to change risk factors which are accessible and likely to make a difference: for example, habits of thinking ('cognition') and patterns of behaviour which can be altered to bring about better results for the individual. The process of change is often seen primarily as the acquisition of new skills. Such approaches are also consistent with the style of work advocated by McGuire and Priestley in Britain, but a particularly influential development in Canada and elsewhere was the idea of a 'programme' which put together a series of carefully planned and designed learning opportunities into a cumulative sequence, covering an appropriate curriculum of skills and allowing plenty of opportunity to reinforce learning through structured practice. (This last aspect was often overlooked by UK practitioners of 'social skills' approaches in probation – see Hudson, 1988.) Robert Ross, for example, after carrying out research which identified a focus on thinking as a common feature of many successful interventions with offenders (Ross and Fabiano, 1985), developed a programme called 'Reasoning and Rehabilitation' which systematically adopted a cognitive-behavioural focus (Ross *et al.*, 1986) and was to exercise a widespread influence on work with offenders both in prisons and in the community.

Systematic research reviews and meta-analysis

The second major strand of research which helped to revive rehabilitation as a feasible goal in criminal justice was a series of research reviews which tried to pull together the findings of what was becoming a substantial body of research, in order to draw out general lessons about what approaches were likely to be effective. Some of these were carried out at the request of governments and were traditional narrative reviews which summarized a number of studies and pointed to shared or important findings: for example, McLaren (1992) in New Zealand and McIvor (1990) in Scotland. The latter was a particularly impressive piece of work and was destined to influence developments throughout Britain. In general, the narrative research reviews of this period found more studies with positive outcomes than had been available to earlier reviewers such as Lipton *et al.* (1975).

In addition to this traditional style of review, researchers and practitioners were also beginning to benefit from the new statistical technique of meta-analysis which combines the results from a number of studies by coding them to a common framework and applying a common measure of outcome. Initially the most influential of the meta-analytic reviews were those of Andrews and his colleagues in Canada (Andrews *et al.*, 1990), and of Lipsey in the United States, which mainly concerned work with young offenders (Lipsey, 1992). Some examples of meta-analysis have been subject to criticisms concerning, for example, the risk of subjective judgment in the coding of studies, the small results base for some conclusions and the possible bias resulting from the greater probability of positive findings being accepted for publication (for these and other problems see Mair, 1994). However, the broad overall conclusions seem soundly based, particularly if we consider how large a number of studies with adverse findings would be needed to overturn them. Similar conclusions have been reached by other substantial and rigorous reviews of the crime reduction literature (including Gaes *et al.*, 1999 and Sherman *et al.*, 1998), and the extreme scepticism still shown by some commentators (for example Mair, 2004) seems unfounded.

In meta-analytic reviews, the impact of methods or approaches on offenders is typically expressed as an 'effect size' which indicates the difference in reconviction rates (or sometimes some other measure) between those offenders who have experienced particular methods or services and those who have had some other input, or no input at all, depending on the particular study design. 'Effect sizes' can be expressed in a number of ways, including correlation coefficients, odds ratios, 'Binomial Effect Size Display' (BESD), or (in the case of outcomes such as change in test scores) the mean change divided by the standard deviation of the initial scores. However, those which are most interesting for our current purposes concern reductions in reconviction which are greater than the reductions which would have been produced either by the customary methods or by doing nothing.

A recent comprehensive overview of 30 meta-analytic reviews published between 1985 and 2001 (McGuire, 2002, p. 13) points out that 'the impact of "treatment" that can be defined in numerous ways is, on average, positive' but 'the mean effect taken across a broad spectrum of treatment or intervention types is relatively modest'. McGuire summarizes this as a difference in reconviction rates of 9 per cent or 10 per cent in favour of those receiving 'treatment'. Such differences, although 'modest', have considerable policy significance if they can be achieved consistently in a field accustomed to results showing no difference. When types of intervention are restricted to those thought most likely to be useful, effect sizes tend to rise: for example, a meta-analysis of 68 studies of the effectiveness of cognitive-behavioural methods with offenders, published at the same time as McGuire's review and so not included in it, shows an average effect size approaching 13 per cent (Lipton *et al.*, 2002). Using the BESD convention this is approximately equivalent to the difference between a 44 per cent reconviction rate in a 'treatment' group and a 56 per cent reconviction rate in a comparison group. Another review of a range of effective projects in Europe reports a 21 per cent difference in reoffending (measured in various ways) between intervention groups and comparison groups (Redondo *et al.*, 2002). Lipsey and Wilson (1998) report some even larger effect sizes from effective work with young offenders.

Meta-analysis in this field has also been concerned to establish not only whether appropriate work with offenders typically has an impact on future offending, but (more ambitiously) to discover what approaches and methods typically produce good results. It is important to recognise that finding an association between particular methods and good results does not in itself demonstrate a causal relationship, and attempts to maximise effects by putting a number of probably effective ingredients together rest at best on a plausible hypothesis which itself requires further evaluation. Nevertheless, throughout the 1990s efforts have been made to draw up lists of the characteristics of effective rehabilitative efforts and to use them as a basis for planning services. For example, the Correctional Services Accreditation Panel (CSAP), which approves programmes and systems of services for implementation with offenders in prisons and the community in England and Wales, requires applicants to have at least a plausible evidence-based hypothesis and a reasonable plan to test it if they cannot point to positive results already achieved in pilot studies (see Rex *et al.*, 2003).

Lists of the characteristics of successful programmes have been produced and updated by a number of commentators, particularly by Andrews and by McGuire, and the latest version (McGuire, 2002, p. 24, drawing on Andrews, 2001) points to eighteen 'principles of effective interventions to reduce recidivism'. These can be summarised and in some cases grouped together, hopefully without too much oversimplification, in the following eleven approaches to design and delivery:

1. Using human service strategies based on 'personality and social learning' theories and on evidence about factors which increase the risk of offending;
2. Using community-based settings or, if in custody, making services as community-oriented as possible;
3. Using risk levels and criminogenic needs, assessed by properly validated methods, to inform targeting and allocation to services;
4. Using multi-modal approaches which match services to learning styles, motivation and aptitude;
5. Adapting services to difference and diversity, and recognizing participants' strengths;
6. Monitoring continuity of services and care, including relapse prevention;
7. Giving staff clear guidance on principles and on where they can use discretion;
8. Monitoring and maintaining programme integrity, i.e. that services are delivered as intended;
9. Developing staff skills, including the capacity to maintain 'high quality interpersonal relationships';
10. Ensuring good knowledgeable management; and
11. Adapting services to local context, client groups and resources.

Such lists are, of course, easier to draw up than to embody consistently in service designs. However, it is also striking to see how closely this recent list resembles similar lists drawn up nearly ten years ago (for example McGuire, 1995; Raynor, 1996). The message from research has been consistent for some time: the most obvious differences in the new list are a stronger focus on the need for practitioners to use interpersonal skills and exercise some discretion, on the need to take diversity among participants into account and on the importance of the broader service context in supporting effective intervention.

At this point it is necessary to record some cautionary notes about how this kind of research is often interpreted. First, it does not offer a guaranteed recipe for success. Lipsey (1999) points to the difference between 'demonstration' and 'practical' interventions. The former are the special pilot projects which are often the source of the research reviewed, and the latter are the routine implementations which follow organisational decisions to adopt new methods. Better results are more commonly found among the 'demonstration' projects: in Lipsey's study the 196 'practical' programmes reviewed were on average half as effective as the 205 'demonstration' programmes. Even this level of effectiveness depended heavily on a few programmes, as 57 per cent of the 'practical' programmes had no appreciable effect. As he points out, 'rehabilitative programmes of a practical "real world" sort clearly can be effective; the challenge is to design and implement them so that they, in fact, are effective' (Lipsey, 1999, p. 641). Other researchers have recently been drawing attention to the crucial importance of implementation, described as 'the forgotten issue in effective correctional treatment' (Gendreau *et al.*, 1999; see also Bernfeld *et al.*, 2001). Some studies (for example Raynor and Vanstone, 2001) have pointed to the particular context of some successful interventions, including enthusiastic practitioners, a culture of curiosity about results and a management style which openly debates principles and methods and encourages staff to own them.[5] Not even the most optimistic senior manager would claim these are always present.

Local evaluations of structured programmes

The third major strand of research which prepared the way for the 'What Works' movement of the 1990s was a small group of studies which provided reasonably convincing evidence for reductions in reconviction among fairly high-risk probationers who had, as part of their probation orders, participated in structured programmes of various kinds designed to address their offending. Such studies were a rarity in Britain after a decade of discouragement, but a few researchers had not completely accepted the 'nothing works' agenda and had the opportunity to carry out evaluative studies with local probation services. Two studies which have already been mentioned in this chapter were published in the late 1980s and showed positive results. The first of these (Raynor, 1988), carried out in South Wales, showed a group of young adult male probationers achieving a reconviction rate some 13 per cent below comparable offenders sentenced to custody, as well as reporting a reduction in social and personal problems. There was also evidence that the project had reduced the use of custodial sentences by local courts. The second study (Roberts, 1989), carried out in Hereford and Worcester, also showed substantial reductions in offending by young adult probationers, including not only reductions in the number offending but reductions in the frequency of offending by those who did. Around the same time, those who actually looked at research from other countries (not a widespread habit in British probation at the time) could study the first comparative evaluation of the Reasoning and Rehabilitation programme (Ross *et al.*, 1988) which showed particularly encouraging results. Other research such as the evaluations of American experiments with 'Intensive Supervision' also contained lessons for those who were interested in effectiveness (for example, Petersilia, 1990) although the overall results were less encouraging.

The first fully evaluated attempt in England and Wales to apply these principles to a programme for offenders supervised by the probation service was started in the (then) Mid Glamorgan probation service in South Wales in 1990, under the leadership of the late David Sutton as Chief Probation Officer. It was known as Straight Thinking On Probation or STOP, a version of Ross's Reasoning and Rehabilitation (R & R) programme. The R & R programme included modules on problem-solving, social skills, management of emotions, negotiation skills, critical reasoning, creative thinking and values enhancement. In Mid Glamorgan these were delivered over 35 two-hour sessions. The evaluation study's findings concerning the programme's impact on offenders have been widely discussed in Britain, largely because of the shortage of other comparable studies at the time. As a consequence, the results have often been quoted as lending support to cognitive-behavioural methods of supervision, and their impact may even appear disproportionate for what are in reality fairly modest outcomes from a local study carried out with small numbers (655 offenders altogether, including 59 programme completers and several comparison groups), little research funding and, at the beginning, very little official encouragement at national level.

For a full account of the findings, readers should refer to Raynor and Vanstone (1996, 1997) and Raynor (1998). Overall, a fair summary of the findings of the STOP evaluation is that there was some evidence of fairly short-term reductions in offending (35 per cent of programme completers reconvicted in a year, compared to a predicted rate of 42 per cent; in contrast, a custodially sentenced comparison group with the same predicted reconviction rate showed 49 per cent reconvicted within a year of release). There were also more persistent reductions in more serious offending among those who completed the programme. These were associated with reported changes in attitudes, thinking and behaviour consistent with the rationale of a cognitive-behavioural programme. On the whole, STOP appeared to offer a more effective and constructive sentencing option than other likely sentences for this group of relatively serious and persistent offenders. However, the findings also pointed to a need to improve the matching of offenders to the programme and the proportion completing it: some programme members were clearly selected on a tariff basis, being at high risk of a custodial sentence, rather than on the basis of assessed needs appropriate to the programme. Most importantly, the study pointed to a need to reinforce what was learned during the programme by appropriate follow-up during the remainder of the period of supervision.

The implementation of the new approaches in Britain

While this and other local experiments were proceeding, managers, practitioners and researchers who were interested in the practical implications of new ideas about effective practice were organising an annual series of 'What Works' conferences to disseminate the new ideas. A number of papers from the first three of these were eventually published in 1995 as a very influential collection edited by James McGuire. A conference organised by Colin Roberts at Green College, Oxford, also helped to promote the new approaches, and in 1993 the Home Office organised a conference in Bath, followed by another conference in London in 1995 on 'Managing What Works'. This was followed by a circular (Home Office, 1995c) encouraging (or requiring) probation services to adopt effective methods and promising follow-up action by the independent Probation Inspectorate (although, as many pointed out at the time, the relevant evidence base for Britain was at that stage quite small).

Different countries and agencies took rather different routes to the implementation of the new ideas. In Scotland, with different criminal justice legislation, there was a tradition of welfare-centred juvenile justice and a criminal justice social work service provided by local authority social work departments rather than a separate probation service. Consequently, the chosen development strategy emphasized education and incremental development in a context where implementation was necessarily devolved and localised. The Scottish Office (later, after political devolution, Scottish Executive) funded an advanced University course for senior practitioners and a Development Unit, and aimed to influence service providers in the right direction by using its powers to set standards and fund services. In England and Wales (the policies come from England, since criminal justice powers are not yet devolved to the Welsh Assembly Government) rather different approaches emerged for young offenders, under the auspices of the Youth Justice Board, and for adult offenders, under the probation service and the prison service. The Youth Justice Board, working through Youth Offending Teams (YOTs) in each locality, encouraged experimentation and diversity by funding a wide variety of local schemes. This was probably a good way of engaging the energies and creativity of local agencies and practitioners, but it created problems for research and for consistency of practice.

The probation service adopted, by contrast, a highly centralised development strategy accompanied by a systematic programme of research. Involvement of the Inspectorate, and of the then Chief Inspector Graham Smith, proved to be an essential catalyst in taking forward the 'What Works' agenda (or, as it was known at first, the Effective Practice Initiative). Instead of a simple inspection to follow up the 1995 circular, a research exercise was set up involving a detailed survey of probation areas by Andrew Underdown, a senior probation manager who was already closely involved in issues around effective practice. The results, eventually published in 1998 (Underdown, 1998), were an eye-opener: of the 267 programmes which probation areas claimed they were running based on effective practice principles, evidence of effectiveness based on reasonably convincing evaluation was available only for four (one of which was not actually included in the responses to the initial survey). One of these was the Mid-Glamorgan STOP programme; the others were in London, where John Wilkinson played an important role in programme evaluation (Wilkinson, 1997, 1998).

These very poor results pointed to the need for a centrally-managed initiative to introduce more effective forms of supervision. The Home Office's Probation Unit worked closely with the Inspectorate to develop what was renamed the 'What Works' initiative; good publications were issued to promote awareness (Chapman and Hough, 1998; McGuire, 2000) and a number of promising programmes were identified for piloting and evaluation as 'pathfinder' programmes, with support in due course from the Government's Crime Reduction Programme (CRP).[6] The pathfinders not only included several cognitive-behavioural programmes (one of them designed by James McGuire, and another being a revised version of Reasoning and Rehabilitation) but also included work on basic skills (improving literacy and numeracy to improve chances of employment), pro-social approaches to supervision in community service, and a number of joint projects run by probation services with prisons and in some cases voluntary organisations working on the resettlement of short-term prisoners after release.

In the meantime a new probation service was taking shape, to come formally into existence as the National Probation Service for England and Wales in April 2001, replacing the old separate area probation services and explicitly committed to public protection and crime reduction. Instead of 54 separate probation services, each responsible to and employed by a local Probation Committee consisting largely of local magistrates, the new National Probation Service was a single organization run by a Director with a substantial central staff located in the Home Office (the National Probation Directorate). Some local influence was still provided by the 42 Area Boards, each employing the staff in its own area (apart from the area's Chief Officer) but responsibility for policy moved to the centre and was implemented through a national management structure. This included 'regional managers' employed by the National Directorate rather than by the Boards in their region. The new areas were coterminous with police, court and crown prosecution service areas in order to facilitate multi-agency working in the criminal justice system (though they did not coincide with local authorities or with Youth Offending Teams), and board members were chosen on the basis of relevant expertise, with much less representation of sentencers than on the old Committees. The new Service started with an annual budget of about £500 million (roughly 4 per cent of overall spending on the criminal justice system) and a staff of around 13,000, of whom about half were probation officers.

This new structure emerged from a substantial review of prison and probation services (Home Office, 1998) which, among other possibilities, considered merging prisons and probation into a single correctional service, but eventually concluded that this would be a step too far. The main aim of the changes was to create an organization which could be more effectively managed and directed from the centre, so that central policy initiatives could be implemented without local priorities or variations leading to the kind of uneven implementation documented by Underdown (1998). Detailed national policies and targets were published (National Probation Service, 2001) incorporating 'stretch objectives' designed to produce change, and performance was monitored. All this represented a considerable transformation over a very short period of time, and the new organization was faced with the problem of how to maintain a sense of involvement among those groups which had less influence in the new structure than they had in the past. These groups included the magistrates who passed most of the community sentences, and some of the Service's own staff. The response of the probation officers' association to the What Works agenda is discussed later in the chapter.

The results of the pathfinders and other British studies

The pathfinder studies were the first concentrated and targeted official research on the effectiveness of the probation service's work with offenders since the mid-1970s, and by far the largest body of research on this subject ever undertaken in Britain. At the time of writing all of them had submitted final or at least interim reports, and although some will not yield all their results until reconviction studies and in some cases 'Phase 2' studies are published, enough work has been done to form at least a preliminary picture of the lessons which need to be learned.

The studies of offending behaviour programmes are being carried out by Leicester and Liverpool Universities, and published findings include a report on implementation issues (Hollin *et al.*, 2002) and a reconviction study (Hollin *et al.*, 2004). The reconviction study covers the three high-volume 'general offending behaviour' programmes which were at the heart of the What Works initiative ('Think First', 'Reasoning and Rehabilitation' and 'Enhanced Thinking Skills') together with a fairly small number of offenders drawn from the 'Priestley One-to-One' programme and a programme for substance abusers, 'Addressing Substance-Related Offending'. Attrition rates were high, both through incomplete data and through failure to complete programmes, with completion rates in the high-volume programmes ranging from 38 per cent to 21 per cent. As a result, the eventual analysis covers 2230 offenders allocated to programmes and a comparison group of 2645 probationers without programme requirements, but only 748 of the programme group actually completed programmes. In what is now becoming a familiar pattern of findings, the programme completers, as a group, were reconvicted less than the comparison group, but the non-completers were reconvicted more than the comparison group. With such low completion rates, the poor results for the non-completers more than cancelled out the good results for the completers, so that it was impossible to demonstrate any net reductions in re-offending resulting from the introduction of the programmes. Also, as discussed later in the chapter, low completion rates make it difficult or impossible to determine how far the benefits to completers are due to the programme, and how far to 'selection effects' resulting from the fact that the people less likely to benefit are also less likely to complete.

Detailed outcome data are also available from another study of the 'Think First' programme, which began to be used in probation settings in 1997 and has been evaluated by the Oxford Probation Studies Unit. This work includes a retrospective study of offenders sentenced to the programme in 1997–98 and a prospective study of offenders sentenced in 2000–2001. Although the full report has not been published at the time of writing, some results have appeared in various sources (for example, Ong *et al.*, 2003; Roberts, 2004). Again reconviction rates for programme completers were significantly better than for non-completers, but completion rates were very low at only 28 per cent. Lower risk offenders were more likely to complete, but their reconviction levels, already low, did not improve. The indications were that better completion rates could be achieved by better targeting, by better case management to motivate offenders, support them through the programme and help with other problems in their lives, and by better follow-up to encourage use of skills learned on the programme.

Although the retrospective study (Ong *et al.*, 2003) found reconviction rates for programme members to be slightly *worse* than for a custodial comparison group, fully reliable comparisons with the outcomes of other sentences cannot be undertaken until a properly matched comparison group can be created using centrally held data. However, the study already has important implications for targeting. For example, the policy of recruiting as many offenders as possible to programmes to meet Treasury targets has probably tended to undermine the fit between offenders' needs and programmes and contributed to increasing attrition and non-completion, which in turn reduces the overall impact of the programme if non-completers reconvict more. A difference in reconviction rates in favour of completers is also found in the most recent evaluation of prison-based cognitive skills programmes (Cann *et al.*, 2003); in this study, completers also performed better than comparison groups. Of the two earlier evaluations of the prison-based programmes, the first also showed significant positive results (Friendship *et al.*, 2002), while the second showed no significant differences, possibly reflecting difficulties in establishing a properly matched comparison group (Falshaw *et al.*, 2003). In prison, of course, completion rates are much higher.

The Basic Skills pathfinder was also evaluated by the Probation Studies Unit (McMahon *et al.*, 2004). Reported findings include the fact that very few of the many offenders with basic skills needs actually started on projects which were intended to help with basic skills, so that the number of 'completers' was so small that little useful outcome data could be collected. Of 1003 offenders assessed as having basic skills needs, only 20 remained in the project long enough to be available for interview after training. This study probably needs to be repeated in a context of much better designed service provision: as it stands it tells us more about the extent of implementation problems than about the likely benefit of basic skills inputs.

The community punishment pathfinder, evaluated by a team led by the Cambridge Institute of Criminology (Rex and Gelsthorpe, 2002; Rex *et al.*, 2004), aimed to add a rehabilitative component to the existing reparative focus of community punishment. The evaluation covered a number of projects which enhanced standard community service provision through pro-social modelling and skills training. Significant gains were seen in crime-prone attitudes, self-reported problems and accredited work-related skills, though some of these were also found in the comparison areas. The reconviction study is still in progress, but few differences in outcome measures are apparent so far between the experimental and the comparison areas. The probation service is now implementing a related initiative called Enhanced Community Punishment which was intended to be subject to further evaluation, as required by the Correctional Services Accreditation Panel, but at the time of writing the planned evaluation has been cancelled on financial grounds.

The resettlement pathfinders for short-term prisoners (Lewis *et al.*, 2003) were evaluated by a team from Bristol, Cardiff and Swansea Universities. This study also showed a number of implementation problems, getting off to a slow start and failing to meet target numbers. Awaiting publication at the time of writing is a second phase study covering a smaller number of projects which aim to apply the lessons of the first evaluation, and a reconviction study is also under way. However, the published study points to some successes: the take-up of post-release assistance was substantially increased by these projects, and participants showed significant positive change in crime-prone attitudes and self-reported problems. What worked best appeared to be a combination of facilitating access to resources relevant to prisoners' needs and taking some steps to address their thinking and motivation, particularly through a short cognitive-motivational group programme undertaken before release.[7]

In spite of differences in focus and findings, some themes emerge consistently from these studies. One is implementation: in all cases the experimental projects did not proceed exactly as planned, with knock-on effects on the research designs. The typical result is not confirmation of what works, but rather a better-informed second phase pilot with a further evaluation. Designs are quasi-experimental with no random allocation, and there are problems in identifying appropriately matched comparison groups, partly due to the fact that although a nationally standardised system for risk and need assessment is being implemented (OASys Development Team, 2001) it was not in place to support the earlier development of 'what works'. Consequently the matching of 'treatment' and comparison groups has tended to be on the basis of criminal history only, and is vulnerable to selection effects arising from differences in need. There are also doubts about whether the quality of reconviction information available from central databases will be good enough to show the fairly modest differences in reconviction rates which would probably be the most that we can expect from these projects (Merrington and Stanley, 2000). Nevertheless, the accumulation of a large body of competent research on the current operation and effectiveness of these initiatives is a major step forward in probation research, on which much future work can be built.

The wide range of innovations sponsored by the Youth Justice Board has proved even more difficult to evaluate, largely because many of them looked for local research strategies to be in place before setting up teams of 'national evaluators', and the national evaluators then found little consistency in the scale, methodology, competence, interest in collaboration or in some cases even existence of the local evaluation arrangements. This made it difficult for them to undertake their national task of coordination and collation of findings (Wilcox, 2003). Some positive findings have been published, for example in relation to the outcomes of such new practices as final warnings (Hine and Celnick, 2001), though without clear evidence that the 'interventions' which accompanied some final warnings improved their effectiveness. There are also some encouraging studies of the implementation of new orders (Holdaway *et al.*, 2001; Ghate and Ramella, 2002; Newburn *et al.*, 2002), but overall there is limited information so far available concerning the effectiveness of rehabilitative and reintegrative disposals promoted by the Youth Justice Board.

Research which has been carried out in the Youth Justice sector on rehabilitative efforts such as cognitive-behavioural programmes (Feilzer *et al.*, 2004) and substance abuse programmes (Hammersley *et al.*, 2004) has found it difficult to draw general conclusions, for reasons similar to those described by Wilcox (2003). High attrition and the difficulty of finding appropriate comparison groups have also created problems in assessing the effectiveness of projects, but the reports contain many lessons about implementation. The final evaluation of the Intensive Supervision and Surveillance Programme is not published at the time of writing, but few specialists in the field expect it to report a straight-forward success. One interesting example of reintegration is the Youth Inclusion Programme, which particularly targets 'at risk' 13–16-year-old young people in deprived areas (Morgan Harris Burrows, 2003). The projects aim to involve young people in a range of constructive activities as an early intervention strategy. There are indications of substantial reductions in arrest rates among those involved, but reductions also occur in a comparison group and it is difficult to disentangle selection and maturation effects from programme effects.

Nevertheless, this inclusive approach to work with young people appears promising, not least because it takes a developmental approach and a positive view of their potential. These characteristics are shared with a particularly interesting project in Scotland, the Freagarrach Project (Lobley *et al.*, 2001) which works intensively with young people who already have substantial histories of offending, using a mixture of activities, programme-like groupwork and mobilization of assistance from other agencies. The research shows clearly the importance of skilled case management, commitment to the young people and their futures, maintenance of strong and effective working relationships with other agencies, effective engagement of young people in the project and persistence through periods of difficulty – an example of the inclusive approach to young people in trouble which also informs other aspects of the Scottish criminal justice system. The available information on reconvictions looks promising but not conclusive, again because of difficulties in establishing an appropriate comparison group. Some other studies from Scotland also show promising results (for example, the Airborne Initiative, which combined adventurous activity with a short offending behaviour programme: McIvor *et al.*, 2000).

Among the more promising developments of this period should be counted the cognitive-behavioural programmes for sex offenders which aimed to challenge the cognitive distortions which supported continued offending, to promote awareness of the harm done to victims, and to teach offenders to recognise when they were at risk of re-offending and to take action to reduce the risk. Such programmes were the focus of a major development and evaluation effort in the prison service (Beech *et al.*, 1999) and showed some positive outcomes as measured by psychometric testing. Modestly encouraging outcomes were also found in some community-based programmes (Becket *et al.*, 1994). Although reconviction studies are a difficult evaluation technique to use with this group due to low reporting and prosecution rates for this type of offence and often low numbers in both 'treatment' and comparison groups, some studies have shown differences in favour of those who have completed appropriate programmes (Allam, 1998, shows non-significant differences, probably due to low numbers; Friendship *et al.*, 2003, found small but significant differences). In addition, some small studies using reasonably convincing quasi-experimental methods have proved encouraging and supported further developmental work on programmes of work with particular groups of offenders (for example, an evaluation of Aggression Replacement Training in the community: Sugg, 2000).

Overall, however, it is difficult to avoid the conclusion of Merrington and Stanley (2004) that 'it is too early to say what works, what doesn't and what is promising' (Merrington and Stanley, 2004, pp. 17–18), at least as far as Britain is concerned. We have certainly learned a good deal about the difficulties of implementation. However, the overall impact of the 'What Works' initiatives is bound to appear disappointing to those leaders and managers of services who invested so much in them. This is particularly true of the probation service's pathfinder programmes (Raynor, 2004), which were unprecedented in scale, implemented in the politicised context of the CRP and carried the additional burden of trying to re-establish the reputation of a probation service which had been the target of political hostility and neglect. If the pathfinders failed, successes elsewhere would not count for much. More than five years have now passed since the pathfinders were established, and we have not so much an end product as a variety of interesting interim products, with a mixture of positive and negative findings and few clear answers to the central questions which underpinned the inclusion of the pathfinders in the CRP. Overall, it is fair to conclude that the pathfinder studies carried out in the probation service have not delivered the unambiguous endorsement of the methods and processes of the 'What Works' project which its leaders, drawing on the international research, originally expected. Seen in the light of the official perception that success needed to be visible within three years, this can easily be seen as failure, and many discussions (for example in the Correctional Services Accreditation Panel) have begun to ask whether we should see this as failure of theory, implementation or research. The remainder of this chapter considers these possibilities, returning at the end to the important question of whether it is yet appropriate to talk of failure at all.

Was the theory wrong?

We have already described some of the international research which was available in the early and mid-1990s which supported the launch of the 'what works' initiative. More recent reviews and meta-analyses cover many more studies, and the conclusions are broadly similar (see, for example, Gaes, 1999; Lipton *et al.*, 2002; McGuire, 2002; Redondo *et al.*, 2002). Cognitive-behavioural methods based on social learning theory, often but not always delivered in group programmes, continue to emerge as a good option for helping some offenders to help themselves to stop offending. Where there has been some shift in the emphasis of recent research it has been in its increased recognition of the critical role played by sound implementation (Gendreau *et al.*, 1999; Bernfeld *et al.*, 2001) and by the personal skills of practitioners (Trotter, 1993; Dowden and Andrews, 2004). Increasing care is also being taken to recognise and control the element of judgment which is inevitably present in coding for meta-analysis, and experienced research reviewers (for example McGuire, 2002) warn against simplistic attempts to use the results of meta-analysis as recipes for programme development.

It is also becoming clear that the weight of evidence in support of cognitive-behavioural methods results partly from the fact that they have been subject to more systematic research than other approaches because they were among the first interventions with offenders that allowed researchers a degree of certainty about the content of services. This permitted a greater clarity than usual about what practitioners were doing, and therefore about the inputs associated with any observed outcomes. Increasingly it is argued that other areas of practice deserve similarly systematic attention. In short, the approaches which have become characteristic of the 'What Works' movement do not claim to be a panacea (they never did), but they continue to enjoy large and growing empirical support from internationally gathered evidence.

However, it is always important to consider the evidence thoroughly, and in the real world many of the managers and practitioners who drove forward the British version of 'What Works' will not have had time to do this. Relying on summaries or extracts creates the risk that important qualifications, limitations and contextual issues will be missed, and this may help to explain some of the problems. In particular, looking only at headline results will often mean that lessons about implementation are overlooked, and this is the next set of issues to consider.

Uneven implementation and 'programme fetishism'

The 'What Works' probation experiment in England and Wales was the largest initiative of its kind anywhere in the world, and the expectation

(built into the CRP) that the major elements of it would be implemented and show results in three years was not realistic. The pressure
created by this time-scale led to short-cuts, some of which had little
evidence available to support them at the time and have in retrospect
proved damaging. For example, the targets for accredited programme
completions set in 1999, which drove the pace of the roll-out of offending
behaviour programmes, were negotiated with Treasury officials without
any systematic prior assessment of the characteristics of offenders under
supervision and their suitability for programmes (Raynor, 2003).

This raises an important point related to risk assessment and 'actuarial
justice'. It is a commonplace of the international literature on effective
rehabilitation that programmes need to be provided for the right offenders.
Not only is it important to observe the 'risk principle' (Andrews *et al.*,
1990) which requires the selection of offenders with a significant risk
of re-offending, there has been a long-standing concern that intensive
intervention can be harmful to those who do not present a significant
risk (Thorpe *et al.*, 1980; Walker *et al.*, 1981). The theories of 'actuarial
justice' reviewed in the previous chapter present this as a question of
efficiency, but in the 'what works' literature it is a question of effectiveness:
selecting the offenders who need the programme is as important as
having a well-designed programme.

Probation officers in England and Wales have been required to include
assessments of risk in pre-sentence reports since the publication of the
first 'National Standards' (Home Office, 1992), but at that time they had
no recognised or reliable technique for doing so. The requirement arose
partly from concerns about 'dangerous' offenders, and discussions of
risk in British probation have always tended to confuse the concepts of
risk of reconviction (how likely is another conviction within a specified
time?) and risk of serious harm or dangerousness. In the latter case, the
priority is to manage the risk and protect potential victims, and knowing
what kind of serious harm might be done in what circumstances to what
kind of victim, and following what kind of changes or trigger events, is
more important than knowing exactly how probable the dangerous
event is.[8] However, questions of appropriate selection for the normal
range of offending behaviour programmes are more likely to depend on
assessing risk of reconviction, and on whether offenders have the particular
needs which the programme seeks to address.

Techniques of risk assessment in this field have been developing
rapidly in recent years (Robinson, 2002). Bonta (1996) describes three
'generations' of risk assessment: the first was the unsupported and
unstructured judgment of individual practitioners; the second was the

development of actuarial scales which used studies of large groups of offenders to calculate a risk of reconviction based on 'static' risk factors such as sex, age and previous convictions; and the third, known as risk/ need assessment, adds the use of assessments of current circumstances, attitudes and needs. Such third-generation instruments are likely to be constructed using actuarial methods, but because they include material derived from structured and systematic assessment of need they can also be used to identify targets for change: in other words, if you can identify the needs which are increasing the risk, and can then meet some of them, the levels of both assessed and actual risk should be reduced. Such methods are designed to support rehabilitation by identifying what kinds of help might assist an offender in reducing his or her chances of reconviction. Theories of 'actuarial justice' appear to be informed by knowledge of second-generation risk assessment methods only: the third-generation risk/need methods are designed to support rehabilitative intervention rather than simply to allocate offenders to graduated levels of surveillance or coercion in response to assessed risk, which is the only function envisaged for risk assessment in 'actuarial justice' (see Robinson, 1999).

Bonta's 'generations' metaphor can usefully be applied to developments in England and Wales. During the mid-1990s a good second-generation instrument was developed, the Offender Group Reconviction Scale (OGRS), derived originally from regression analysis of the criminal records of 13,711 offenders (Copas, 1992), and considerably refined and updated since. This was widely used in research and evaluation, as well as by probation officers preparing pre-sentence reports. Also in the mid-1990s risk/need instruments were introduced in some areas: these were the ACE instrument (Assessment, Case-management and Evaluation) developed in England, and the internationally used Canadian instrument LSI-R (Level of Service Inventory – Revised). Although the Home Office initially showed little understanding of risk/need assessment, eventually a study was commissioned based on ACE and LSI-R (Raynor *et al.*, 2000), which showed that they provided both a useful degree of risk assessment and an assessment of needs, and that they could also be used to measure changes in needs during supervision which correlated with changes in the risk of reconviction.

Instead of adopting one or both of these demonstrably practical and reasonably effective ready-made instruments, the Home Office had by then decided to develop its own. The circular announcing this was issued, perhaps deliberately, just days before submission of an interim report on ACE and LSI-R in 1999 which demonstrated the validity of both. Ignoring international experience which suggested that the development of a really user-friendly and practical instrument could take ten years or more, the new Offender Assessment System was to be rolled out to both prisons and probation in 2000. Unfortunately the development process took far longer than this, and the design which emerged was scientifically impressive but cumbersome to use, proving initially unpopular with many staff. The result is that it is still not in comprehensive use at the time of writing, and the two services have yet to reap the benefits of a third-generation instrument. In other jurisdictions such instruments (particularly the LSI-R) have been used for some years to support evidence-led targeting strategies and service evaluation (see, for example, Miles and Raynor, 2004).

The implications of this for pathfinders were that risk/need methods were not used to inform the setting of target numbers, and most of the selection of offenders for programmes was based either on the need to fill programmes and achieve numerical targets, or at best on targeting a medium-high risk group measured using OGRS. Pathfinder evaluations (Hollin *et al.*, 2004) have shown that many offenders were selected from outside the target group, but that those within it were more likely to do well. It seems at least possible that if developments had proceeded in a more logical order, with assessment methods in place before programme targets were set, the targets might have been a more realistic assessment of need, the roll-out of programmes would have been slower, the people in them would have been selected better, the completion rates would have been higher and the overall result better.

Other implementation difficulties were frequently noticed during the pathfinder research process. As well as the additional stresses and complexities created by the change to a National Probation Service in 2001, researchers noted a number of problems within the pathfinder projects themselves. Projects were often not running in a fully developed form when the evidence which would be used to measure their effectiveness was collected. Often, as in the resettlement study, local projects depended on small numbers of staff and were vulnerable to staff sickness or communication problems (Lewis *et al.*, 2003). In all the pathfinder studies, projects tended to make a slow start and not to achieve their target numbers; in the 'basic skills' and 'employment' pathfinders (Haslewood-Pocsik *et al.*, 2004; McMahon *et al.*, 2004) numbers completing were so small that the evaluation could not be carried out as intended, and in the 'offending behaviour' pathfinders (for example Roberts, 2004) the high levels of attrition led to difficulties in interpreting evidence: outcomes based on a small number of completers may be effects of the programme or may be simply the effects of whatever selection or self-selection processes led to those people, rather than others, completing the programme. In general the better a project is implemented, the easier it is to draw conclusions from evaluating it.

In addition, the top-down management style which was seen as necessary to drive implementation forward within the prescribed time-scale (Blumsom, 2004) alienated parts of the workforce, particularly probation officers who were used to a high degree of autonomy. Staff in some areas found their workloads spiralling out of control at the same time as demands to meet targets were increasing. At one stage (luckily after most of the pathfinder data had been gathered) most probation areas were involved in industrial action over workloads, and the probation officers' union NAPO expressed its concern in conference resolutions which rejected aspects of the 'What Works' approach (NAPO, 2001). In such circumstances researchers could hardly be surprised if some of the data quality was poor. (This and other implementation difficulties are, of course, consistent with the findings of Lipsey's 1999 meta-analysis of 'demonstration' and 'practical' projects discussed earlier in this chapter.)

Some other practices which were introduced at the insistence of politicians rather than by Service managers seem also to have made implementation more difficult. Tougher enforcement and more breaches were seen as a political imperative throughout the 1990s, and the reduction of officers' discretion, although producing a more consistent approach, also helped to increase the number who failed to complete programmes. No independent research has fully addressed the consequences of current approaches to enforcement, though recent Home Office research has shown no beneficial effects on offenders (Hearnden and Millie, 2003) and several commentators have questioned whether some of its effects are counterproductive (Ellis, 2000; Hedderman and Hearnden, 2001). In addition, the priority given to 'programmes' by the CRP seems to have contributed to a tendency to be preoccupied with implementing programmes or 'interventions' rather than with providing an experience of supervision which would be effective as a whole. Although this had been pointed out by early British research (Raynor and Vanstone, 1997), by the Chief Inspector of Probation (Professor Rod Morgan) when he warned against 'programme fetishism' (Her Majesty's Inspectorate of Probation, 2002) and by the Correctional Services Accreditation Panel which insisted on continuity as one of its accreditation criteria (Correctional Services Accreditation Panel, 2003), little attention was paid to the need for effective case management until attrition rates started to cause concern. Recent Home Office research (Partridge, 2004) has begun to examine the merits of different case management models, but meanwhile research in other countries has pointed clearly to the benefits of continuity of contact with skilled practitioners (Trotter, 1993; Dowden and Andrews, 2004). The pathfinder projects were not designed with this in mind, and the associated evaluations are therefore able to say little about its contribution to outcomes, which may have been considerable.

Another frequent criticism of the 'What Works' initiative has been its slowness to address issues of diversity in the offending population, and particularly the needs of women offenders and minority ethnic offenders (Kemshall *et al.*, 2004). Some critics have argued that both the cognitive-behavioural approach and the risk assessment techniques which have underpinned the 'What Works' movement are developed from research on majority correctional populations of white male offenders, and are therefore inapplicable to women and ethnic minorities (Shaw and Hannah-Moffatt, 2000, 2004). Others have taken the view that social learning theory is likely to have applications beyond the white male population, but that diversity of needs and experiences, together with the structurally disadvantaged position of non-dominant social groups, requires specific investigation and tailor-made or modified approaches (for example, Porporino *et al.*, 2003; Calverley *et al.*, 2004). What is, however, clear is that although British probation services had a long history of attempts to address specific needs of women and of Black and Asian offenders, pressure to deliver numbers in the context of the CRP ensured that major development efforts were initially concentrated on rolling out programmes designed for large numbers of offenders (Raynor, 2003). Efforts to develop and evaluate provision for minorities[9] came later, leading for example to a women's programme (Porporino *et al.*, 2003) and to a variety of 'pathfinder' projects for Black and Asian offenders. It is too soon to tell how successful these will be, though some implementation problems are again being noted (Lovbakke and Homes, 2004; Stephens *et al.*, 2004).

Lessons for research

One further explanation advanced for the 'failure' of the pathfinders to demonstrate success has been, in effect, that they were evaluated using the wrong kind of research. Recent thinking within the Home Office (reflected, for example, in a recent research review [Harper and Chitty, 2004] and in internal guides to research standards [Research, Development, and Statistics, 2004a,b]) suggests that the results would have been more conclusive if Randomised Control Trials (RCTs) had been used, and that more of these should be used in the future. Such thinking is clearly influenced by the 'Maryland scale' (Sherman *et al.*, 2002) and other attempts to define a hierarchy of reliability in research designs. The pathfinder evaluation process was undertaken in the hope of finding statistically significant differences in outcome between an 'intervention' and its absence. RCTs clearly offer the most persuasive way of demonstrating such differences, and criminal justice research in Britain has suffered as a result of the rarity of RCTs, but it would be unwise to put all our heuristic eggs in this one basket, just as it was unwise to concentrate development efforts almost exclusively on 'programmes'.

Unless we know in detail how outcomes are produced we are unlikely to be able to replicate them (which is, after all, the point of the enterprise). Just as the drive for 'programmes' overshadowed significant contributors to effectiveness such as case management, practitioner skills and indeed sentencing patterns, so a one-dimensional approach to outcome measurement is in danger of concentrating on the dependent variable without sufficiently exploring what all the independent variables might be. A method which was developed for hypothesis-testing may not be the best way of understanding a process: that requires a more descriptive exploration, and the use of a variety of methods and viewpoints (as advocated, for example, in pluralistic evaluation [Smith and Cantley, 1984]). Moving straight to the hypothesis-testing stage without going through a process of exploration, description and understanding runs the risk of testing the wrong hypothesis. As Hedderman puts it, 'the likelihood that a programme alone would be effective may have been overestimated, making a holistic approach seem less essential' (Hedderman, 2004). In practice many of the pathfinder evaluation reports used a much broader model of social research, but there was always a sense that official interest centred on only one kind of narrowly defined 'result'. It is encouraging that there is now more recognition of the need for broader research designs, including longitudinal designs which track offenders through the correctional system (Hedderman, 2004). However, the pressure on criminal justice researchers to provide quick, simple and convenient answers will remain as long as criminal justice policy-making continues to be subject to the pressures of populism and electoral anxiety, and there seems little prospect of an end to these. Even the Home Office's own internal evaluation of the whole CRP points to the difficulties created by its high political profile (Homel *et al.*, 2005). One of its conclusions states:

> the review clearly identified that the CRP's problems were not the result of any intrinsic failure of this [evidence-based] policy approach.

Rather they stemmed from a lack of understanding of (or commitment to) what is really required to implement such an approach. (Homel *etal.*, 2005, p. xiii)

Forgetting the lessons of history

One further striking feature of the whole pathfinder process has been its tendency to be very selective in its use of past experience and research. This shows itself particularly in two ways. First, there has been a tendency to see effective practice as coterminous with group programmes. This is related to the problem of 'programme fetishism' described above: it is like concentrating on the ball and ignoring the pitch. In addition, it diverts attention from other forms of evidence-based practice: not only programmes work. Second, there has been a failure to locate the 'what works' efforts in their penological context. What sentencing pattern was meant to result from all this effort? We have already mentioned how lessons drawn from the pursuit of 'alternatives to custody' were forgotten when the political wind blew in a different direction. There was also a failure to learn some of the lessons of earlier 'what works' research.

To deal first with the need to look beyond group programmes, we need to recognize that the fact that so much research has concentrated on programmes does not mean that there is no evidence about anything else. Even within the programme paradigm groups are not the only delivery option. McGuire defines a programme simply as a 'structured sequence of opportunities for learning and change' (McGuire, 2002, p. 27), while the definition used by the Correctional Services Accreditation Panel for England and Wales is 'a systematic, reproducible set of activities in which offenders can participate' (CSAP, 2003, p. 25).

Over-preoccupation with group programmes runs the particular risk of neglecting practitioner skills in the case management and supervision process. There is a substantial research literature concerned with effective practice in psychotherapy and social work, some of which would have been familiar to probation officers trained in England and Wales before the separation of probation officer training from social work training in 1997, and should still be covered in the training of probation and criminal justice staff who gain social work qualifications in other jurisdictions. Particular areas of interest here include core facilitative or therapeutic skills, widely researched in the 1960s (Truax and Carkhuff, 1967), which include empathy, positive regard or concern, 'genuineness', and a concrete and specific approach to goals, expectations and processes.

Similar issues continue to be identified in more recent research, and are brought together by McGuire (2003) in a recent discussion of the need for a 'working alliance' rather than a coercive or confrontational relationship. These are not woolly aspirations but concrete skills which are strongly supported by evidence and can be enhanced by training (for an example from social work education see Raynor and Vanstone, 1984). There is also a useful body of research on the enhancement of motivation to change (Miller and Rollnick, 1992) by using skilled interviewing to increase awareness of a need to change and willingness to do so. Much of the evidence here comes from the field of substance abuse, but there are increasing indications of the relevance of motivational work with offenders (for example, Harper and Hardy, 2000).

Other useful components for the development of a 'what works' approach to individual supervision and case management include the practice of 'pro-social modelling', applied to probation practice in Australia by Trotter (1993, 2001) and taken up more recently in a number of British projects (Rex and Matravers, 1998). In Trotter's formulation this involves both the modelling of pro-social attitudes and behaviour by staff supervising offenders, and the acknowledgment and rewarding of such behaviour on the part of offenders themselves. Early research indicated that supervision by officers trained in this approach resulted in lower reconviction rates (Trotter, 1993). While official attempts to build on these findings are now beginning in Britain (for example in the new Enhanced Community Punishment scheme) other forms of practice with some empirical support are not much discussed, perhaps because they are seen as belonging to the social work tradition rather than the correctional field. These include, for example, 'task-centred casework' (Reid and Epstein, 1972), a highly focussed approach to identifying problems, reaching agreements about them, sharing responsibility for addressing them and evaluating outcomes. This approach suggests a number of interesting starting-points for thinking about case management, and was even evaluated in a probation setting with interesting results (Goldberg *et al.*, 1985); however, there has been little sign of any recent attempts to build on these.

In a recent article Dowden and Andrews (2004) report on a meta-analysis of the contribution of certain staff skills to the effectiveness of rehabilitative work with offenders. They define these skills as 'Core Correctional Practices' or CCPs, which can be summarised briefly as effective use of authority; appropriate modelling and reinforcement, the use of a problem-solving approach, and the development of relationships characterised by openness, warmth, empathy, enthusiasm, directiveness and structure. The mean effect sizes of programmes were found to be higher when these were present, and significantly higher when other principles of programme effectiveness were also applied: staff skills and programme design complemented each other, rather than one being a substitute for the other. However, the authors point out that 'Clearly these CCPs were rarely used in the human service programs that were surveyed in this meta-analysis . . . These results suggest that the emphasis placed on developing and utilizing appropriate staff techniques has been sorely lacking within correctional treatment programmes' (p. 209).

Turning now to the question of penological context, we have seen how some of the impetus behind the shift towards evidence-based practice in Britain came from early local studies which showed modest positive effects from various forms of special programme for young adult offenders (Raynor, 1988; Roberts, 1989). These studies were carried out at a time when young adult offenders were increasingly receiving custodial sentences, after which their reconviction rates were particularly high. The studies were therefore located within the policy context of alternatives to custody, and the special forms of supervision they offered were specifically targeted on young people at significant risk of receiving a custodial sentence and unlikely to be made subject to a standard probation order without special content and requirements. These programmes were therefore a success if they recruited young people who would otherwise receive custodial sentences, and if they achieved better results than the custodial sentences they replaced. One study (Raynor, 1988) used similar young people receiving custodial sentences as the comparison group, and also documented local reductions in custodial sentencing during the life of the project. Comparisons with the outcomes of ordinary probation orders were not a major issue because there was plenty of evidence that the project participants would not have received ordinary probation orders. Thus development of such projects made sense as an alternative to custodial sentences, but a different set of questions would have been raised if they had been developed as alternatives to simpler forms of probation order. Also, the very high reconviction rates of young adult offenders following custodial sentences make it easier for non-custodial programmes to out-perform them. It is hazardous to use such studies to promote the advantages of structured and demanding interventions *without* considering what would otherwise be happening to the offenders concerned.

Some years later, the findings of the STOP experiment (Raynor and Vanstone, 1996) were, as we have seen, among those which provided

support for the feasibility and value of programmes in British probation. However, it is important to recognise that the initial planning of
the STOP experiment began in 1990, in a service still aiming at diversion from custody, and eligibility for the programme was determined in
part by an offender's risk of receiving a custodial sentence as well as by
a fairly rudimentary assessment of needs. When the courts did not
agree with a probation officer's recommendation that a particular
offender should undertake the programme, the result was almost always
a custodial sentence, and actuarial prediction of expected reconviction
rates confirmed that programme participants were a high risk group
comparable to those receiving custodial sentences rather than to those
receiving standard probation orders. The beneficial effects of the
programme on reconviction were, as usual, fairly modest, and might
not have attracted much attention in the 'nothing works' era; however,
what is important for current purposes is that those benefits were
mainly demonstrated by comparisons with the effects of custodial
sentences. The advantages over conventional probation orders were less
substantial, but, as in the studies mentioned above, the target group
would have been unlikely to receive such orders in any case.

The point of exploring this in some detail is to remind readers that
the early evidence of successful programmes for relatively high risk
offenders subject to community sentences in England and Wales was
gathered in studies which were primarily looking for appropriate ways
to enhance probation orders to make them marketable and effective for
those otherwise at risk of custodial sentences. They were not primarily
about better ways of supervising people who would get probation
orders anyway; nor were they about substituting programmes for the
ordinary processes of supervision by a probation officer. The programmes
were an additional element in a probation order, not the whole probation order in themselves. The neglect of these issues in the initial planning
and rushed execution of the pathfinders is perhaps understandable.
What is more surprising is to see that both issues (non-programme
forms of effective practice, and the penological context of early 'what
works' research) continue to be ignored in the Home Office's attempt to
review what has been learned from the pathfinders (Harper and Chitty,
2004). However, and in spite of these reservations, probably the most
important legacy of the pathfinders lies in what has been learned from
them. Not only have they generated the largest body of research on
what actually happens when our correctional services make a large
investment in rehabilitation, they have also resulted in a large minority
of probation officers and a significant number of prison staff being

trained in the concepts and methods of evidence-based effective practice, a situation which would have been unthinkable only ten years ago. The process of building these lessons into theory and practice is just beginning.

Critical criminology has, so far, largely failed to engage with the debates and evidence surrounding effective rehabilitation. Attempts to help offenders to stop offending are glibly dismissed as 'controlling' or 'correctional', and little attempt is made to distinguish between pure coercion and the collaborative learning processes on which successful rehabilitation depends. For example, Feeley and Simon comment generally on community sentences:

> Thus, community-based sanctions can be understood in terms of risk management rather than rehabilitative or correctional aspirations. Rather than instruments of reintegrating offenders into the community, they function as mechanisms to maintain control, often through frequent drug testing, over low-risk offenders for whom the more secure forms of custody are judged too expensive and unnecessary. (Feeley and Simon, 1992, p. 461)

Others dismiss the knowledge-base of effective rehabilitation as positivist or reductionist (Kendall, 2004), or as insufficiently informed by a social-structural critique of contemporary society (NAPO, 2001). We hope that this chapter has shown that the real story is more complicated than this, and that the thinking and evidence behind contemporary efforts are both sophisticated and developing. The last ten years have seen a mixture of successes and failures, and an even larger volume of inconclusive outcomes: the process of judging what is working, what is promising and what would be better abandoned will continue for years, and conclusions drawn at this point are necessarily provisional. However, our own provisional conclusion at this stage would be that the 'What Works' movement in Britain tried to move too fast too soon: the correctional services, under considerable political pressure, tried to do many of the right things but were often not able to do them in the right way. Performance in some areas is likely to improve as lessons are learned, but one of the central goals of the 'what works' initiatives, namely the 'rehabilitation' of the probation service itself, was not achieved, and the Service seems likely to pay the price through further reorganisation, fragmentation and perhaps the loss of its name and identity (see NAPO, 2005).

This chapter, then, has told a mixed story about the mainstream developments which have led the attempt to make rehabilitation more effective. In the next chapter we turn to some other developments which have not been so clearly located in the mainstream of policy, but which offer some important alternative approaches whose potential has yet to be fully developed.

7
Against the Tide: Non-treatment Paradigms for the Twenty-first Century

In this chapter we continue our consideration of contemporary rehabilitation, but here we move away from the more explicitly 'correctional' modes of intervention which were considered in Chapter 6. In the present chapter we turn our attention to what might be termed a number of 'non-treatment paradigms': that is, contemporary developments in theory and practice which do not fit within the correctional mould, but which nonetheless clearly belong under the umbrella of rehabilitative initiatives.

In the first part of the chapter we consider the contribution of restorative justice. Although restorative approaches are often associated with the needs and rights of victims of crime, in this chapter we restrict our discussion to the consideration of restorative justice as a possible route towards offender rehabilitation. We then consider recent developments in theory and research around desistance: that is, 'naturalistic' rehabilitation, which we touched upon in Chapter 1. Finally we turn our attention to contemporary attempts to secure the reintegration of a specific group of offenders, namely released prisoners. We conclude the chapter with a summary of what we think these approaches share in common, and what they have to offer those engaged in rehabilitative work with offenders.

Restorative approaches

'Restorative justice' is a term which has been used to denote a wide variety of practices, carried out in a number of different social, cultural and historical contexts. Indeed, it is perhaps best described as a 'conceptual umbrella' under which a number of different practices have found common ground. At its broadest, restorative justice denotes a strategy or set of strategies oriented towards the resolution of conflicts or disputes, not necessarily acts defined as criminal offences. Thus, it has been deployed to deal with anything from school bullying to the resolution of political conflicts (e.g. Dignan and Lowey, 2000). In the more specific context of offending, restorative justice is perhaps best described as an approach which seeks to deal with or respond to offending by involving both offender and victim, and sometimes members of the wider community. This 'essence' of restorative justice is neatly captured in Marshall's popular definition:

> Restorative Justice is a process whereby parties with a stake in a specific offence collectively resolve how to deal with the aftermath of the offence and its implications for the future. (1999, p. 5)

The most common operational examples of restorative justice are *victim-offender mediation*, and *restorative conferencing*. In victim-offender mediation, contact between offender and victim is facilitated by a specially trained, neutral third party (mediator or facilitator). Contact between the two parties may be direct (i.e. face to face) or indirect, involving the relaying of questions and/or information by a mediator/facilitator. The Victim/Offender Reconciliation Program (VORP), which began life in the mid-1970s in Kitchener, Ontario, is generally recognised as the first victim/offender mediation scheme bringing convicted offenders and their victims face to face; whilst the first 'diversion' scheme utilising mediation dates back to 1971 in Columbus, Ohio (Wright, 1991). Recent research indicates that victim/offender mediation is currently the most common form of restorative justice practice in both the United States and Europe. Restorative conferencing differs from victim/offender mediation principally in that it tends to involve members of the wider community as well as the victim and offender. Dignan (2005) distinguishes between two main variants: family group conferencing, and police-led community conferencing. *Family group conferencing* originated in New Zealand as a means of dealing with offending by young people, and as an antidote to criminal justice processes which tended to offer little opportunity for victim involvement, and which were perceived as potentially discriminatory in respect of the Maori population (Maxwell and Morris, 1993). *Police-led community conferencing*, in contrast, originated in the early 1990s in the small town of Wagga Wagga in New South Wales, Australia. Subsequently, this particular conferencing model spread not just to other parts of Australia, but also to the USA, and the UK.

Restorative justice, retribution and rehabilitation

As is well known, the relationship(s) between restorative justice and (retributive) criminal justice have been the subject of intense debate for a number of years (e.g. von Hirsch *et al.*, 2003). Whilst some proponents have conceived or understood restorative justice as an alternative to and as a critique of retributive criminal justice, placing restorative justice in the role of a radical 'alternative paradigm' (e.g. Zehr, 1990), others have conceived restorative justice in 'separatist' or 'reformist' terms (see Dignan, 2005, p. 106). In many of these formulations retributive justice has been criticised for its obsession with establishing guilt and apportioning blame, for the socially excluding stigma of guilt and criminal punishment, and for its failure to meet the needs of victims of crime.

Much less discussed is restorative justice's relationship with notions of rehabilitation. However, certainly in some writing the two are contrasted, or viewed in oppositional terms. For example Braithwaite explains that restorative justice is most commonly defined in terms of 'what it is an alternative to', namely retribution and rehabilitation (1999, p. 4). An earlier example of the restorative/rehabilitative contrast is provided in the writing of Albert Eglash (1977), who is commonly credited with coining the term 'restorative justice'. Eglash suggested that restorative justice stood in contrast to both retributive justice (based on punishment) and distributive justice (based on the therapeutic treatment of offenders). Both the punishment and the treatment models, he argued, share a focus on the actions of offenders, deny meaningful victim participation in the justice process and require merely passive participation on the part of the offender. Similarly, Schiff has argued:

> As a justice strategy, restorative justice is concerned with much more than simply what is done to or with offenders, and as such it is a much more ambitious justice response [than] either retribution, deterrence, rehabilitation or incapacitation, all of which have far more modest, offender-centred goals. (2003, p. 330)

Certainly there is some validity to such arguments (although we would not necessarily agree that rehabilitation implies 'modest' goals). For example, restorative justice differs from most rehabilitative approaches by virtue of its attempts to involve and meet the needs of victims, and sometimes other stakeholders. Indeed, the current restorative justice 'boom' is commonly understood in the context of a 'victim's movement' which has sought over a number of years to raise the profile of crime victims, highlight their needs and promote their involvement in justice processes in a number of ways (e.g. Miers, 2004; see also Dignan, 2005). In some accounts of restorative justice, whole communities are implicated. So, it seems fair to say that restorative justice is not generally considered to be an offender-centred approach.

Nonetheless, there are strong grounds to argue that the distinctions between restorative justice and rehabilitation have tended to be overplayed. On a conceptual level, we saw in Chapter 1 that the notion of 'restoration' is central to most definitions of rehabilitation. Connections between the two are also acknowledged in the work of a number of restorative justice advocates. For example, having argued that restorative justice stands in contrast to both retribution and 'treatment', Eglash (1977, p. 99) goes on to say that, for him, restorative justice is primarily about 'justice and rehabilitation for offenders'. Similarly, Braithwaite's work is quite explicitly concerned with rehabilitative concepts, not least the notion of offender reintegration, which we shall explore more fully below. Indeed, Braithwaite makes it clear that he sees restorative justice as a means of reducing re-offending, and oriented ultimately towards the goal of crime prevention. In this vein he understands 'restoring offenders' in terms of reduced offending, and in one well-known paper engages in a lengthy discussion about why restorative justice practices 'rehabilitate better than criminal justice practices grounded in the welfare model' (1999, pp. 27, 67–69). This point about Braithwaite's work has also been noted by Dignan (2005, pp. 102–05), who argues that Braithwaite's theory of reintegrative shaming is primarily offender focused and conceived as offering a more effective means of controlling crime rather than principally constituting a means of meeting the needs of victims.[1] It is also worth noting that in 2001, the Home Office decided to provide funding to three restorative justice schemes under the auspices of its Crime Reduction Programme[2] (Shapland *et al.*, 2004). Indeed, restorative justice interventions are increasingly being evaluated in terms of reconviction, which seems to indicate a growing interest in the capacity of restorative justice interventions to promote desistance from offending (e.g. Sherman, Strang and Woods, 2000; Miers *et al.*, 2001; Wilcox, Young and Hoyle, 2004).

Johnstone (2002) agrees that restorative justice is concerned with reform or rehabilitation. He argues that the distinction which is commonly drawn between restorative and rehabilitative approaches is based on a 'caricature' of the latter, which assumes, first, that the offender is a 'passive object' in need of expert intervention; and, secondly, 'a highly "medicalised" model of penal treatment, in which experts attempt to cure offenders of their criminal tendencies through psychiatric treatment and other techniques' (2002, p. 111). This image of 'rehabilitation-as-treatment' arguably exaggerates the distinction between the two types of approach, and tends to mask the fact that restorative justice *does* concern itself with offender rehabilitation – albeit that the rehabilitation of offenders is considered a legitimate 'end' of restorative justice 'only insofar as it can be made compatible with the goal of achieving justice for their victims' (2002, p. 95).

Restorative justice's concern with offender rehabilitation is perhaps best exemplified in the work of Gordon Bazemore. Bazemore's work is important because he takes care to distinguish between the 'treatment model', on the one hand, and the pursuit of rehabilitative outcomes on the other. For example, writing about juvenile justice in the US, Bazemore has argued that the main problem with the treatment model lies in its 'singular focus on the psychological needs and social deficits of the offender' which, he argues, portrays the offender as an 'object of remedial services or therapeutic intervention' (1996, p. 48). Elsewhere Bazemore (1998) highlights the treatment model's tendency to ignore the 'relational' or interpersonal domain which he, in common with other restorative justice advocates, sees as central to effective rehabilitation. These characteristics, he argues, explain why a treatment-oriented approach has done so little to promote – and indeed has been an obstacle to – a *reintegrative* response to juvenile offenders. From the purview of restorative justice, Bazemore argues, the offender is a truly active subject: a community member capable of making amends to victims and/or communities, and acquiring new competencies in the process. As such, restorative justice offers a 'broader reintegrative agenda' which is at once 'more empowering, effective and marketable' than the treatment model (1996, pp. 52, 42). Bazemore's arguments in favour of restorative justice are thus informed by a critique of the 'treatment model', but he also acknowledges and indeed emphasises the capacity of restorative justice to achieve offender rehabilitation – or, more specifically, reintegration (see further below).

Restorative justice and rehabilitation may, then, have more in common than is often assumed or acknowledged. But although arguably overlapping rather than distinct approaches, restorative justice offers a perspective – or set of perspectives – on the rehabilitation of offenders which sets it apart from the 'new rehabilitationism' described in the previous chapter.

Shaming and the 'moral discourse' thesis

Although not all restorative justice advocates view 'shaming' as a necessary ingredient of restorative justice (e.g. Maxwell and Morris, 2004), ideas and arguments about shame and shaming are nonetheless important when it comes to understanding the rehabilitative potential of restorative interventions. Braithwaite's (1989) theory of reintegrative shaming grew out of a critique of retributive criminal justice. This critique holds that the public shaming of offenders which characterises criminal justice is negative and destructive, and generates a stigma which the offender finds difficult to shake off. 'Stigmatic' shaming, it is argued, is negative because it is associated with social exclusion, and/or the acceptance by the offender of a criminal identity. Both are likely to lead to further offending.

At the heart of Braithwaite's theory is the idea that shaming need not be stigmatic or counter-productive and can, in fact, be an effective means of eliciting remorse on the part of the offender, as well as a precursor to forgiveness, acceptance and reintegration within the law-abiding community. To quote Braithwaite:

> Stigmatization is shaming which creates outcasts, where 'criminal' becomes a master status trait that drives out all other identities, shaming where bonds of respect with the offender are not sustained. Reintegrative shaming, in contrast, is disapproval dispensed within an ongoing relationship with the offender based on respect, shaming which focuses on the evil of the deed rather than on the offender as an irreclaimably evil person, where degradation ceremonies are followed by ceremonies to decertify deviance, where forgiveness, apology, and repentance are culturally important. (1993, p. 1)

So-called 'reintegrative shaming' relies not on formal criminal justice personnel, but rather the active participation of the victim(s) and a number of 'significant others' whom the offender respects. The role of the victim in the shaming process involves drawing attention to the 'collateral damage' caused by the offence, which may include fear, personal injury and/or material loss (Braithwaite and Mugford, 1994). Faced with the victim's personal testimony, it is argued, the offender is less likely to be able to employ 'techniques of neutralization' (Sykes and Matza, 1957) which serve to minimise the harm caused by offending, and more likely to have to face up to the consequences of his or her actions. The offender's 'significant others' are also important in that their disapproval of the offence ('moral censure') is likely to mean more to the offender than that of a magistrate or judge, in whose esteem the offender has little or no personal investment (Braithwaite, 1999, pp. 39–40). As Johnstone (2002, Chapter 5) explains, in restorative justice the community's 'judgement' is not simply a prelude to pain (punishment), but rather is intended to perform an educative and reintegrative function: the concern is to persuade offenders to share the community's judgement of their behaviour. For this to happen the judgement must not be unduly harsh, and must be 'mixed with empathy for [offenders] as members of the community who have erred' (2002, p. 93).[3]

Dignan (2005, p. 102) has recently argued that Braithwaite's theory of reintegrative shaming forms an important part of an emerging 'moral discourse thesis'. Central to this thesis, Dignan argues, is the idea that the offender's conscience is potentially a much more powerful weapon against deviant behaviour than is punishment, and that engaging the offender in 'normative or moralizing dialogue' can be an effective inhibitor of future offending behaviour. Another important proponent of this 'moral discourse thesis', Dignan contends, is Anthony Duff, who has attempted to theorise a 'communicative theory' of punishment. Much like Braithwaite's theory, Duff's focuses on engaging offenders in a moral dialogue and views the censure of offending behaviour as a prelude to contrition and, ultimately, rehabilitation (Duff, 2001, 2003). For Dignan, this emphasis on encouraging a moral perspective constitutes common ground for restorative justice and some contemporary 'correctional' approaches, such as cognitive-behavioural treatment which, as we saw in the previous chapter, encourages offenders to take responsibility for their actions and to acknowledge the harm caused to others.

'Reintegration ceremonies'

As noted in the previous section, Braithwaite views shaming as a prelude to reintegration. Reintegration as a concept is not always particularly well defined in the restorative justice literature (or, indeed, elsewhere), although there are exceptions. For example, Van Ness and Strong explicitly state that 'When we speak of reintegration we mean re-entry into community life as a whole, contributing, productive person' (1997, p. 116). Generally speaking, 'reintegration' is a term which reflects restorative justice's broad concern with the communities in which both offenders and victims live, and which are considered to have a great deal both to offer and to gain from 'restorative' processes.

In the restorative justice literature, reintegration tends to have both symbolic and practical expressions. Symbolically, reintegration can denote the acceptance of the offender as a law-abiding member of the community: his or her 'requalification' as a citizen, also referred to as the 'decertification of deviance' (Braithwaite and Mugford, 1994, p. 141). This may be underlined by the acceptance of an offender's apology, offers of forgiveness, and/or the signing of an agreement. It may also be achieved through the offender's performance of reparative work – which may form part of such an agreement – upon which forgiveness or re-acceptance may be conditional.

For some writers, the performance of reparative work – whether to direct victims or the wider community – can in itself be a *practical* means towards reintegration. Bazemore (1996), for example, writing about juvenile offenders, sees reparation as an opportunity for offenders to be actively engaged – possibly for the first time – in roles that allow them to gain valuable (and valued) skills, and practice 'being competent'. According to this view, reparation enables offenders to see themselves, and to be seen by others, as valuable resources with something to offer the community, rather than passive recipients of 'help' (see also Maruna, 2000, Chapter 6). We shall revisit this so-called 'strengths-based' approach in the following chapter.

Another example of practical reintegration is offered by Braithwaite and Mugford, who talk of 'reforming the deviant through reconstructing his or her social ties' (1994, p. 140). In Braithwaite's work the role of offender supporters is seen as crucial to reintegration: 'the support of those who enjoy the strongest relationships of love and respect with the offender structures reintegration into the ritual' (1999, p. 39). However, Braithwaite also refers to reintegration into 'a wider web of community ties and support' (Braithwaite and Mugford, 1994, p. 139). In a similar vein, Bazemore (1996, 1998) argues that successful offender rehabilitation is dependent upon the active support and involvement of the community.

Restorative and correctional interventions: Happy bedfellows?

It is clear from the preceding section that one of the key characteristics of restorative justice, at least for some of its proponents, is its emphasis on the mobilisation of community resources and support in the process of offender rehabilitation. This begs the question of whether restorative justice sees any role at all for more overtly 'correctional' interventions, and/or rehabilitative 'professionals'. On this point, opinion appears to be divided. For example, in Bazemore's earlier work, he argues that rehabilitation is something which can be effectively devolved to the community, rather than residing with 'experts' (Bazemore, 1996, 1999). Braithwaite (1999, p. 69), on the other hand, does entertain the possibility that restorative justice and 'corrections' can work side by side. Braithwaite argues that restorative justice can be a forum in which the offender's need for mainstream 'rehabilitative programs' (i.e. those managed by 'experts') can be voiced, and in this context he cites the examples of drug treatment, job training, counselling and educational programmes. However, he admits that restorative justice's empowerment of citizens to choose from rehabilitative programmes has tended to be more rhetoric than reality. In a similar vein, Johnstone has argued that 'in many cases, the goal of reintegrating offenders into the law-abiding community has a better chance of being achieved if both therapeutic and restorative interventions are employed, in a coordinated programme, rather than if we rely upon one to the exclusion of the other' (2002, p. 111).

More recently, Bazemore appears to have 'come round' to this more accommodating position. Indeed, in an interesting recent paper, Bazemore and Bell (2004) present a convincing vision of how offender treatment programmes might be made more 'restorative' and, in their view, effective.

> We believe that it is possible to develop a fully restorative model of rehabilitation that would outperform other models. This holistic approach would apply restorative principles to existing treatment programs, to the way decisions are made about obligations, to the way in which offenders are supported and supervised in the community...this approach is based on the premise that rehabilitation is important, but not in isolation from a community or relational context. (2004, p. 120)

Within this proposed model, the role of citizens or community members as 'natural helpers' continues to be emphasised, but a role is also recognised for professional treatment 'as needed'. Such treatment, it is argued, is in turn likely to be enhanced by the application of restorative justice principles. For example, intervention programmes could incorporate such elements as victim awareness and conflict resolution sessions, and/or 'restorative' community service (in which offenders and community members work together on projects that contribute to community well-being). A further suggestion, which to a large extent mirrors some of the arguments put forward by desistance researchers (see below), is a 'strategic extension' of the offender/supervisor relationship (assuming such a relationship exists). Bazemore and Bell argue that treatment professionals could do more to strengthen the offender's relationships with family, friends, peers and supporters – relationships which, it is argued, are both more sustainable and influential than those between professional and 'client'. Continuing with their analysis of 'effective correctional interventions' from a restorative justice perspective, they go on to suggest a broader view of criminogenic need as 'an absence of pro-social relationships, effective guardianship, and social support' (2004, p. 126).

From this perspective, then, restorative justice continues to offer a critique of 'treatment' approaches, but it does not render them irrelevant. Perhaps more importantly, it usefully reminds us that offender *treatment* and offender *rehabilitation* are not coterminous – a point which we stressed in Chapter 1. As Bazemore and Bell (2004, p. 128) put it: 'rehabilitation is bigger than changing the attitudes, thinking, and problematic behaviour of offenders... what good is an initial change in thinking and behaviour patterns if the offender's relationships with law-abiding adults and peers are weak or non-existent?' In our view, this is a point with which it is extremely difficult to argue.

Desistance: Exploring 'naturalistic' reform

> Most people in communities who are familiar with former delinquent offenders such as 'Johnny', a young gang member who 'settled down' at the age of 25, know rehabilitation is often not the result of any complex therapeutic process. Johnny was not 'rehabilitated' because he suddenly stopped using drugs or gained some new insights through counselling, but because an employer, his wife, his uncle, and other adults eventually provided him with a job, family ties, and a network of support. Those who know Johnny best also know that there was most likely a time when his criminal activities began to pose a risk to these emerging adult relationships. (Bazemore, 1999, p. 170)

In Chapter 1 we introduced the notion of 'naturalistic' rehabilitation, noting that whilst the concept of rehabilitation is often associated with the intervention of a third party, it is not a necessary ingredient of the rehabilitative process. Here we explore this model of rehabilitation in a little more depth, and point to a resurgence of interest in the processes through which offenders come to rehabilitate themselves often without the aid or assistance of 'experts'. We also consider how knowledge about natural processes of desistance have been utilised to inform the ways in which offenders are dealt with, particularly by the community-based 'correctional apparatus'.

The 'age-crime curve' is not a recent criminological discovery. As Maruna (2000, p. 10) explains, the relationship between ageing and the abandonment of criminal behaviour is one of the oldest and best-known findings of criminological research, dating back at least to the work of Adolphe Quetelet in the early nineteenth century. However, as a number of researchers have observed, it is also one of the least understood phenomena in the field. Maruna identifies two main types of explanation for the apparently natural cessation of offending among young adults. At the risk of over-simplifying, 'ontogenic' accounts tend to subsume the process of desistance to the biological process of maturation. The Gluecks' (1940) theory of 'maturational reform' exemplifies this tradition, as does the more recent work of Gottfredson and Hirschi (1990, 1995). According to ontogenic explanations, the majority of young offenders simply 'grow out' of crime, with the 'sheer passage of time' converting the offender into a responsible adult (Glueck and Glueck, 1940; cited in Maruna, 2000, p. 27). However, this approach has been criticised for its failure to adequately explain *how* processes of ageing or maturation impact on offending behaviour: it is far from clear whether or how various mechanisms mediate between the two (see Maruna, 2000, p. 30).

Sociogenic accounts, by contrast, emphasise the role of external factors (relationships, employment, education etc.), and are aptly summarised by Maruna under the heading 'a steady job and the love of a good woman' (2000, p. 27). In theoretical terms, life events such as getting a job or entering a relationship are conceived as 'social bonds', offering the offender, perhaps for the first time, a stake in conformity and a reason (or reasons) to 'go straight' (e.g. Hirschi, 1969; Sampson and Laub, 1993). Offenders who fail to establish such bonds, conversely, are prone to continue offending because they have little or nothing to lose. This basic theoretical model is supported by some research which has demonstrated connections between such life events as partnership formation, gaining employment and leaving home, on the one hand, and desistance on the other. However, the research evidence is far from unequivocal: empirical connections have tended to be weak (e.g. see Farrall and Bowling's [1999] review of research). Sociogenic accounts have thus been criticised for over-simplifying the relationships between external/structural factors and desistance, and ignoring the role of human agency in mediating between the two. For example, Farrall and Bowling have suggested, firstly, that offenders are not ' "super-dupes" who react to wider social forces and situations' and, further, that 'the process of desistance is one that is produced through an *interplay* between individual choices, and a range of wider social forces, institutional and societal practices which are beyond the control of the individual' (1999, p. 261; see also Laub and Sampson, 2001).

Growing interest in the role of human agency in desistance – and the interactions between such agency and structural factors – spawned a small crop of research studies during the 1990s, which aimed to explore the 'phenomenology of desistance' (Maruna, 2000, p. 32). In other words, they sought to understand offenders' subjective experiences of 'going straight'. This body of research included Burnett's longitudinal study of 130 persistent property offenders, who were interviewed prior to and following release from prison sentences (Burnett, 1992, 2000; Burnett and Maruna, 2004); as well as Maruna's comparative study in which samples of desisters and active offenders in Liverpool were interviewed (Maruna, 2000). Both studies found that a key difference between 'persisters' and 'desisters' was their sense of personal agency, or perceptions of control over their own future. These studies seemed to show that desistance is not something which 'happens to' offenders independently of their own will to desist.

Burnett conducted interviews with offenders just prior to release from prison, with a maximum of two follow-up interviews in the two years following release. One of the key findings of this study was that offenders' own predictions regarding their chances of desisting were fairly accurate. In other words, there was a strong correlation between self-reported estimates of the likelihood of reoffending and self-reported offending following release from prison. This led Burnett to conclude that offenders' confidence, optimism, or sense of control over their future, could well have played an important role in determining outcomes. In a ten-year reconviction study, Burnett and Maruna (2004) found that although problems with employment, drugs or marital relations as measured at the second interview were predictive of longer-term reconviction, measures of pre-release optimism continued to play a significant role, in that the most optimistic offenders at the start of the study seemed better able to cope with the problems they encountered following release.

In Maruna's study, whilst persisters tended to view themselves as relatively powerless 'victims of circumstance', central to the typical desister's identity was an optimistic perception of personal control over his own destiny. This tended to be coupled with 'generative' goals or plans – that is, an active desire to forge a productive and worthwhile present. 'Whereas active offenders...seemed to have little vision of what the future might hold, desisting interviewees had a plan and were optimistic that they could make it work' (Maruna, 2000, p. 147). Offenders' generative plans often included a desire to 'give something back' to the wider community; a desire highly consistent with the goals of restorative justice, as discussed above.

Applying desistance research to correctional approaches

In an attempt to apply the findings of his study to the work of the correctional services, Maruna has argued that rehabilitative interventions might focus less attention and resources on trying to change committed offenders and more on providing support for those showing evidence of motivation to change and having already embarked on the process of desistance. Maruna also emphasised the role of significant others (particularly figures of authority) in 'recognising redemption' and 'decertifying deviance': that is, acknowledging and confirming to the wider community the offender's successful transition to a crime-free, productive life. Maruna argued persuasively that reformation is:

> a construct that is negotiated through interaction between an individual and significant others in a process of 'looking-glass rehabilitation'. Until ex-offenders are formally and symbolically recognised as 'success stories,' their conversion may remain suspect to significant others, and most importantly to themselves. (2000, p. 158)

In this context Maruna noted the role in the UK of the Rehabilitation of Offenders Act (1974) in 'decertifying deviance', which we discussed in Chapter 1. He also acknowledged the potential contribution of restorative justice, recognising the role of reintegration ceremonies in this context. However, he argued that such positive recognition of rehabilitation is a rarity in contemporary criminal justice systems.[4]

Other researchers, too, were beginning to think about how emerging findings in respect of processes of desistance might inform rehabilitative, or 'correctional', interventions, particularly probation.[5] In Britain, two studies have been particularly influential in this respect. In the first, Rex (1999) examined desistance from the point of view of a group of 60 probationers and their supervising probation officers. Rex was particularly interested in the role probationers ascribed to probation in enabling them to desist from offending. She found that, from the perspective of the probationers interviewed, certain qualities of the supervisory relationship were perceived as more likely to foster, or reinforce, a commitment to desist. Rex summarised this as 'an "engaging" style in which empathy was balanced by a certain amount of professionalism and formality' (1999, p. 372). Half of the probationers Rex interviewed manifested feelings of personal loyalty towards their supervisors, and she suggested that much of this commitment seemed to stem from the support, encouragement and positive reinforcement offered by probation officers. Rex's research clearly highlighted the potential of the supervisory relationship in motivating and assisting offenders in their progression towards desistance – a feature of supervision which had received little attention for some time, despite being a central plank of 'effective' rehabilitation as theorised in earlier decades (e.g. Davies, 1969). It also recalled the classic qualities of the effective counsellor (namely genuineness, warmth and empathy), described by Carl Rogers several decades previously – and familiar to generations of social workers in the UK and elsewhere (Rogers, 1951).

The second study (Farrall, 2002), which also involved interviews with probationers (this time a larger sample of 199), featured a longitudinal design and had the advantage of being able to attend to the correspondence between offenders' statements about their experiences of probation and subsequent desistance (or offending). Farrall's findings seemed to indicate that desistance was largely independent of probation intervention. Instead, key correlates of desistance seemed to be offender motivation and changes in social circumstances (principally, employment and family relationships). Although Farrall did concede that supervision may have been important in encouraging and maintaining motivation, he was critical of probation officers' apparent reluctance to intervene in respect of (primarily social) 'desistance-related' factors.[6] Drawing on Coleman's (1988, 1990) distinction between 'human' and 'social' capital, Farrall concluded that his findings were at odds with probation's contemporary preoccupation with cognitive-behavioural interventions, with their emphasis on building 'human capital' (skills, knowledge etc.). Instead, he argued, probation ought to be doing more to foster offenders' 'social capital' – namely the productive interpersonal and social relationships which facilitate social integration.

Whilst Farrall's call for probation to re-think its '*exclusive* focus on cognitive behavioural work' (2002, p. 228, emphasis in original) is perhaps over-zealous, and erroneously portrays the probation service as ignorant of alternative approaches,[7] his proposal that probation work might benefit from becoming more 'desistance focused' has been well received, and further developed by Fergus McNeill (2003, 2004). Writing from a Scottish perspective, where community-based interventions with offenders continue to be managed and delivered by local authority social workers, McNeill has argued that a prescription for desistance-focused practice recalls many of the traditional concerns – and skills – of social work (abandoned as the basis of probation training for probation officers in England and Wales in the mid-1990s). Drawing on the research by Rex, Farrall and others, McNeill has argued that such a prescription would need to be sensitively individualised, grounded in an effective one-to-one relationship between offender and worker, and sufficiently cognisant of the offender's needs in respect of social capital. The latter element, he argues, forefronts the importance of (social work) skills in advocacy and work with families. Recalling Bottoms and McWilliams's (1979) 'non-treatment paradigm', as well as Drakeford and Vanstone's (1996) *Beyond Offending Behaviour*, McNeill puts forward a convincing case for the revival of 'practical help' as a legitimate part of rehabilitative interventions.[8]

A desistance-focused approach would also appear to recommend greater attention to working with offenders' motivation to change, a point acknowledged by both McNeill and Maruna. There is a useful body of research on the enhancement of motivation to change, derived from the field of substance abuse (Miller and Rollnick, 1992), and 'motivational interviewing', which aims to increase awareness of a need to change and willingness to do so, has been successfully applied to work with offenders (for example, see Harper and Hardy, 2000). However, the extent of its use by probation officers is not known. In a recent study of programme attrition (Kemshall and Canton, 2002, p. 7), motivational interviewing was identified by probation areas as the main mechanism employed to combat attrition, but the researchers were unable to ascertain whether staff were trained and/or competent in this method. So, whilst many staff may be aware of motivational interviewing as a potentially effective method, it appears that there is much room for increasing its prominence and use in the interest of promoting desistance.

Reintegration, resettlement, reentry

We have already encountered the notion of 'reintegration' in the context of both the restorative justice and desistance literatures, in particular, in the work of John Braithwaite. We noted that in Braithwaite's work, reintegration is theorised as an outcome of effective shaming, and is envisaged principally as a process of 'decertifying deviance': that is, re-involving a former offender as a normal citizen in the reciprocal processes of community life. However, we also noted a broader and more practical notion of reintegration, understood as attempts to involve former offenders in networks of pro-social opportunities and relationships with the aim of helping them achieve and/or maintain a non-offending lifestyle. As we have seen, the building of the so-called 'social capital' is not only a central part of some versions of restorative justice, but also central in recent studies of desistance, which emphasise the value of social and community bonds in increasing the former offender's investment in 'going straight'.

Both these meanings are relevant when the notion of reintegration is used to describe the release of prisoners into the community. As we noted in Chapter 1, the theory and practice of (ex) offender reintegration is commonly called 'resettlement'.[9] In the United States the preferred term for this is 'prisoner re-entry' (Maruna and LeBel, 2003; Petersilia, 2003). In both Britain and the United States, the subject of ex-prisoner reintegration is currently receiving renewed attention, both from academic commentators and policy-makers (Maruna *et al.*, 2004), for two main reasons. The first of these is the increase in both countries of numbers of offenders being sentenced to custody, a trend which in England and Wales at least has been shown to be related more to punitive policies than any significant changes in the type or seriousness of offending (Lewis, 2004). The other is the high level of reoffending (and in many cases re-incarceration) among released prisoners (e.g. Travis and Petersilia, 2001). In Britain, the Social Exclusion Unit (2002, p. 5) has reported that of those prisoners released in 1997, 58 per cent were reconvicted within two years, with 36 per cent serving another prison sentence as a result. Among 18–20-year-old males the situation was even more dire: 72 per cent were reconvicted within two years of release, with almost half (47 per cent) receiving a further prison sentence. Clearly, for policy makers keen to make an impact of reoffending rates, this is a growing population which they cannot afford to neglect. Indeed, it is one in respect of which even minor achievements in terms of reductions in reoffending rates would be likely to make a major impact in terms of financial savings and costs, both material and otherwise, to victims of crime.

In both Britain and the United States, the provision of pre-release planning and programmes for prisoners has been criticised as woefully inadequate, and identified as a significant part of the 'problem' of effective resettlement. In the US, this problem has been exacerbated by an increase in the average length of prison sentences, which translates into a longer period of detachment from family and other social networks (Travis and Petersilia, 2001). For Travis and Petersilia, the 'inescapable conclusion is that we have paid a price for prison expansion, namely a decline in preparation for the return to community. There is less treatment, fewer skills, less exposure to the world of work, and less focused attention on planning for a smooth transition to the outside world' (2001, p. 300). In England and Wales, conversely, a trend in favour of more short prison sentences, which offer no statutory period of supervision post-release, has been highlighted as one of the main barriers to successful resettlement (Maguire *et al.*, 2000; HMIP, 2001; Social Exclusion Unit, 2002). These prisoners, serving twelve months or less:

> are likely to serve their sentences in under-resourced, overcrowded, local prisons, and are liable to be moved suddenly away from their home region in order for the prison to accommodate new arrivals and remand prisoners. They are likely to have few opportunities for purposeful work, acquiring educational or work skills, or seeing through offending behaviour programmes. On release, the probation service has no statutory responsibility to deal with them at all, and therefore is unlikely to be able to find the resources to do so. (HMIP, 2001, p. 5)

However, the lack of pre-release planning and preparation is only one part of the problem of resettlement. There is now ample evidence that prisoners are among the most disadvantaged and socially excluded groups in society, and therefore the least likely to be amenable to the 'quick fix' of correctional interventions. Based on its comprehensive analysis of the relevant criminological and social research to date, the Social Exclusion Unit (2002) reported that compared with the general population, prisoners were thirteen times as likely to have been in care as a child, twenty times as likely to have been excluded from school, and thirteen times as likely to be unemployed. In addition extremely poor levels of basic skills were reported, as well as high levels of drug abuse and mental disorder. As if this were not problem enough, it is also well known that a prison sentence can serve to exacerbate many of the problems which those entering prison already face, as well as potentially creating new ones. For example, the Social Exclusion Unit reported that a third of prisoners lose their home whilst in prison, two-thirds lose their job, over a fifth face increased financial problems and over two-fifths lose contact with their family. In the majority of US states a felony conviction carries the additional exclusionary punishment of the loss of voting rights. For many offenders, then, the experience of prison exacerbates an already significant history of social exclusion.

Research conducted in the United States presents another worrying perspective on ex-prisoners and social exclusion, revealing that the growing numbers of released prisoners are increasingly returning to, and thus concentrated in, neighbourhoods already facing significant disadvantages. Citing research by Lynch and Sabol, Travis and Petersilia (2001) explain that of just under half a million state prisoners released in 1996, an estimated two-thirds were released into counties containing the central cities of metropolitan areas – cities which are typically less wealthy than neighbouring areas and facing other challenges such as the loss of labour market share to suburban regions. They add that the 'revolving door' of prison may further destabilise already disadvantaged neighbourhoods by disrupting a community's social network, affecting family formation, reducing informal control of children and income to families, and lessening ties among residents (Clear *et al.*, 2001). Travis and Petersilia conclude that not only are prisons turning offenders out less well prepared for their return to the community but they are returning them to communities that are not well equipped to accept them.

All of these issues lead us to the conclusion that it is in relation to resettlement that arguments about 'state-obligated rehabilitation', discussed in Chapter 2, are most persuasive. It will be recalled that Rotman (1990) contends that imprisonment, for many, incurs harmful effects or side-effects (e.g. loss of employment or home; loss of family ties) which constitute punishment in excess of that intended when a sentence of imprisonment is passed. For this reason, he argues, the prisoner should have a right to a positive programme of rehabilitative action to reinstate him to a former position or status (restoration), as an 'antidote' to imprisonment. Further, if we accept that social circumstances play a part in crime causation by restricting the legitimate opportunities available to individuals, then surely the state ought to assume some responsibility and be prepared to offer rehabilitative opportunities to those who would be less likely to offend as a result? (e.g. Carlen, 1989).

Rethinking reintegration

Whether considered from an economic, social or rights-oriented perspective, it is clear that resettlement policies have, for a number of years, been failing not only offenders, but also their families and the communities to which they return. In short, effective reintegration – however defined – is only rarely being achieved.

One particular problem which has been identified, both in the UK and the US, is a lack of coordination between prison and community-based services and resources. Travis and Petersilia have recently argued that US corrections ought to be much more attuned to the 'mission' of reintegrating prisoners, and appropriately organised and funded to link services 'on both sides of the wall', from job training opportunities to housing to mental health care (2001, p. 308). To aid such a 'seamless transition', they suggest the abolition of the current system of parole in favour of 'a new system focused squarely on the goal of reintegration' (2001, p. 306).

Under their proposed scheme, every substantial period of incarceration would be followed by a period of 'managed reentry', but this would be divorced from the original punishment, such that release from custody would signal that the whole sentence had been served, and non-compliance with post-release supervision would not trigger a return to custody. They do however envisage a system of 'graduated sanctions', up to a short deprivation of liberty, such that the offender's obligations during the reentry phase would not be entirely voluntary.

In England and Wales, renewed interest in 'managed reentry' among policy makers has resulted in significant legislative changes, the majority of which have yet to take effect. Reflecting the key problem of unsupervised discharge from short custodial sentences, these changes extend the notion of the 'seamless sentence' to prison sentences of all lengths. Under the new sentencing structure recommended in the Halliday report (Halliday, 2001) and enacted in the Criminal Justice Act 2003, short licence-free sentences are consigned to history, such that every period in custody will be followed by a period of statutory supervision in the community. Hopes of engendering more 'joined-up working' between prison and probation services, and rendering the so-called 'seamless sentences' a reality for all, are also reflected in the establishment of a National Offender Management Service, which brings the prison and probation services under a single organisational umbrella (Carter, 2003; Home Office, 2004a). Government targets which increase the accountability of both prison and probation services in delivering effective resettlement have also been published (Dalkin and Padel, 2004; Home Office, 2004b). Reflecting the Social Exclusion Unit's empirically informed depiction of prisoners as a population poorly equipped for crime-free living, the Home Office's *Action Plan* includes targets in respect of a wide range of issues including accommodation; education, training and employment; mental and physical health; drugs and alcohol; and financial issues. The role family support can play in assisting prisoners with resettlement, and the support needs of families themselves, also receives some attention – though arguably not enough.

Although yet to be fully implemented, this inevitable move towards compulsory 'aftercare' for all has prompted concerns that we may be heading toward the 'waste management' model of resettlement characteristic of the Californian system of parole (Simon, 1993). There, compulsory post-release supervision has resulted in a dramatic rise in the prison population which is almost entirely a function of revocations for technical violations, such as positive drug tests, rather than new offences (Maruna, 2004). However, other commentators have cautiously welcomed the reforms. For example, Maguire has argued that:

> If the system operates efficiently, and if anything like sufficient resources become available (both, of course, dangerous assumptions to make), this should offer a greater opportunity for creative and effective approaches to the resettlement of short-term prisoners than has ever existed before in the UK. (2004b, p. 22)

Maguire reflects enthusiastically on three recent developments in England and Wales which, he argues, hold out some promise for the seamless sentences of the future: the use of short cognitive/motivational programmes delivered partly in prison and partly after release; the use of community-based 'mentors' to 'advise, assist and befriend' offenders and maintain contact post-release (see also Lewis *et al.*, 2003); and the development and implementation of regional 'resettlement strategies' based on partnerships between key service agencies.

In the United States, the new concept of a 'reentry court' is providing the main context for experimentation in respect of ex-prisoner reintegration. Based on the drug court model (initially established in the US but subsequently piloted in England and Scotland – see Robinson and Dignan, 2004), the reentry court casts sentencers in the role of 'reentry managers' who oversee the offender's return to the community. As Maruna and LeBel (2003) explain, the reentry court experiment is still in its infancy, and pilot sites in various US states differ significantly in their approaches. However, in general terms the reentry court uses the authority of the court to apply graduated sanctions and positive reinforcement, and to marshal resources to support the prisoner's reintegration. The core elements of the model, they explain, include *assessment and strategic reentry planning*, involving the ex-offender, sentencers and other key partners; *regular review meetings*, involving the ex-offender and his circle of supporters; and *accountability to the community* through the involvement of citizen advisory boards and other community groups. Rewards for success include the possibility of early release from parole obligations, and 'graduation ceremonies' similar to those used in drug courts (Maruna and LeBel, 2003, p. 2).

Maruna and LeBel argue that the reentry court represents a promising development: not least because, in contrast to traditional parole, the notion of rewarding success is intrinsic to its design. Indeed, all reentry courts are required to outline milestones in the reentry process (such as the completion of voluntary work) which will automatically trigger recognition and appropriate rewards. As they note, rewarding positive achievements, rather than punishing violations, is an unusual but welcome role for the criminal courts. The reentry court also stands in contrast to traditional parole by virtue of its attempts to involve members of the community in the reintegration process. For Maruna and LeBel, the community, not just the court, is an essential participant in the reentry process: not only must the offender negotiate the *physical* reentry into the community on release; but he must also accomplish some sort of 'relational reintegration' back into the wider *moral* community. This of course implies reacceptance of the offender by the community as a law-abiding citizen. Travis and Petersilia (2001) agree that community involvement in reentry is likely to be key to its effectiveness because it is within the community that the 'positive power' of social networks (i.e. social capital) can be found.

Taking these two roles of the reentry court together (community involvement and rewarding success), Maruna and LeBel argue that the reentry court model could potentially be pushed further in both directions. That is, rather than simply rewarding specific acts, it could 'build gradually to a more holistic 'earned redemption' of the participant's character and reputation' (2003, p. 15), signalling the ex-offender's reacceptance into the community of law-abiding citizens. The ultimate reward, they contend, might be the official 'decertification of deviance': that is, the expiration of the individual's criminal history and permission to 'wipe the slate clean'. Here Maruna clearly draws on the findings of his own desistance research, which revealed the value which many desisters appeared to place on 'redemption rituals' orchestrated by those in positions of authority, such as judges (Maruna, 2000, p. 13).

Non-correctional paradigms: Common themes

In this chapter we have deliberately moved away from the more mainstream 'correctional' model of rehabilitation, and turned our attention instead to three comparatively marginal areas of theory and practice, addressing restorative justice, desistance and resettlement. In attempting to cover a substantial amount of ground, it is likely that we have failed to do justice to any of these areas. Nonetheless, we feel the 'journey' has been worthwhile, and it has served to highlight a number of emerging themes which we would be loath to neglect.

It is hopefully clear that far from representing competing paradigms, the approaches discussed in this chapter share a great deal in common. This is especially evident in that a number of academic commentators are familiar contributors to more than one of these literatures. It is also evident in some of the use of terminology, most notably the concepts of reintegration, social bonds and social capital which crop up repeatedly.

How, then, might we summarise their common ground in the context of the subject of this book? Perhaps most fundamentally, the restorative justice, desistance and resettlement literatures all locate rehabilitation firmly *in its social context*. More specifically, they each suggest that rehabilitation can (and should) be conceived as a *relational process*. At its most basic level, this means that successful rehabilitation is most likely to be achieved in the context of relationships and interactions with others, and that the building or re-building of relationships is a legitimate focus of rehabilitative work (e.g. Bazemore, 1998). They remind us that, to a great extent, rehabilitation is a process which can be achieved, and arguably must be negotiated, with a range of 'stakeholders' which includes families, other supporters, and also members of the wider community – not least employers and, in some cases, victims. This relational context is key even in the desistance literature, which at first glance appears to contradict the notion of 'relational rehabilitation'. For although desistance research is most closely associated with the idea of 'self-determined' rehabilitation, it also highlights the value of significant others in terms of helping the offender to negotiate his or her path towards desistance. Farrall's research in particular highlights the importance of fostering the ex-offender's 'social capital': that is, the productive interpersonal and social relationships which can facilitate social integration. The building of social bonds is also a key focus of restorative justice. As Bazemore expresses it, 'a relational approach to rehabilitation cannot be clinical in its focus' (1998, p. 790). He continues:

> At the individual level, if crime is viewed as the result of weak bonds, a relational rehabilitation must be focused primarily on strengthening the offender's ties or bonds to conventional adults and peers and on changing the offender's view of law-abiding citizens and the community. At the community level, intervention to strengthen bonds must focus on changing citizens' views of offenders and increasing the willingness and capacity of community groups to take responsibility for integration and reintegration. (Bazemore, 1998, p. 787)

Research on desistance adds another dimension to the relational perspective, reminding us that the positive regard and acknowledgement of others can be crucial to the consolidation of a non-criminal identity

(Maruna's research being particularly noteworthy in this respect). 'Relational rehabilitation', then, implies that rehabilitation is best understood as a two-way street, requiring commitment and effort not only on the part of the offender, but also, in a variety of ways, on the part of family members, supporters, and community members.

Closely related to the notion of relational rehabilitation, and a common theme again in all three literatures, is the idea that rehabilitation is a process that can to some degree be *devolved* – to families and other social networks lying outside the domain of the traditional 'correctional services', and indeed to offenders themselves. In the contexts of restorative justice, desistance and resettlement, the notion of devolving rehabilitation does not exactly correspond with the notion of 'responsibilising' offenders which characterises the revival of correctionalism (see previous chapter) – although offenders themselves are certainly afforded a central role in their own rehabilitation, most notably in the desistance literature. Rather, what is implied here is a marked reduction in dependency on correctional interventions in rehabilitative processes, and a corresponding increase in the involvement of communities, defined in various ways. This is a characteristic which sets these approaches quite clearly apart from correctional rehabilitation. For example, in restorative justice, reintegrative shaming relies far less on formal criminal justice personnel than on the active participation of victims and significant others whom the offender respects. Restorative justice also emphasises the mobilisation of community resources and support in rehabilitative processes. In the desistance literature, too, the role of families and wider communities as 'natural helpers' or allies in rehabilitative processes is stressed, and advocates of the new resettlement strategies echo the same point. For example, Travis and Petersilia (2001: 309) argue that responsibility for reintegration should not rest solely with correctional agencies, and that 'reentry activities' should take place as close as possible to the local communities in which offenders will ultimately return. It is in these local communities, rather than in correctional agencies, they argue, that the 'positive power of social networks' can be found and exploited.[10] As we have already noted, this notion of 'social capital' is a key one, utilised in all three literatures to describe the ways in which communities may offer resources and opportunities to offenders as they negotiate their journeys towards desistance.

There are of course some problems with the notion of devolving responsibility for offender rehabilitation to families and communities, which should not be overlooked. In particular, there is a tendency in much of the literature discussed in this chapter to idealise the communities in which offenders reside, and to assume access to resourceful social networks, including supportive families. Yet we cannot assume that offenders will inevitably have access to such sources of informal support. The same can be said of 'communities' more generally. This is made particularly clear in some of the US 'reentry' research, which has highlighted the problems of social exclusion that many of the neighbourhoods to which released prisoners return face. Writing on the subject of restorative justice, Van Ness and Strong (1997, p. 121) have argued that individualistic and urbanised Western communities pose potential problems for offender reintegration. They ask whether it is wise to always assume that there are people able and willing to do the challenging work of reintegration. If opportunities to develop social capital are indeed restricted, then clearly the scope for achieving effective reintegration is limited. In this context, economic as well as social opportunities are clearly also key. If employment is one of the primary mechanisms through which ex-offenders negotiate reentry and/or desistance, then depressed economic conditions, or limited economic opportunities in particular community contexts, will surely limit the scope for successful rehabilitation.

This line of argument leads us to the suggestion that some of the resources currently being channelled into the 'correctional services' might be more effectively utilised in the arena of community work, an area of practice from which the probation service, at least in England and Wales, has been gradually distancing itself since the 1980s, largely due to the imposition of nationally defined objectives and priorities (Home Office, 1984). This is not to say, however, that the approaches considered in this chapter render correctional interventions or personnel redundant. Notwithstanding a clear emphasis on the devolution of rehabilitation, and a tendency to de-emphasise the role of experts, the approaches discussed in this chapter can be – and indeed have been – usefully applied to the work of those engaged in 'correctional' work with offenders. We have for example seen some of the ways in which advocates of restorative justice have attempted to marry restorative and correctional approaches. We have also seen, most clearly in McNeill's work, that some of the findings of research on desistance have already been applied to the work of those responsible for supervising offenders in the community.

In fact, we can observe in the field of corrections a marked revival of interest in what we have loosely termed 'relational rehabilitation'; a revival which owes a great deal to the growing literature, and empirical research, associated with the approaches discussed in this chapter (Robinson and Dignan, 2004; Robinson, 2005). The rediscovery of relational rehabilitation is being felt in a number of areas, not least in the context of community service work, where there have been interesting attempts to build on earlier promising results from Australia (Trotter, 1996, 1999; Rex and Matravers, 1998). Interest in how different models of case management impact on offenders is also growing, and is set to increase in the context of the National Offender Management Service in England and Wales (Partridge, 2004). Relational considerations are also emerging in the arena of sentencing and 'sentence management', where a number of new initiatives are encouraging communication, both between sentencers and offenders and between sentencers and those responsible for implementing community-based sentences (Robinson and Dignan, 2004). Attention to the ways in which different parts of the criminal justice and correctional apparatuses relate to one another has come to focus in particular on resettlement, where a lack of coordination and communication has been identified as a significant barrier to 'seamless' reintegration, and a variety of new initiatives are being put in place in an effort to remedy this situation.

There are, thus, a number of ways in which the notion of relational rehabilitation can inform – and indeed is informing – aspects of correctional practice. Far from rendering correctional approaches redundant, theory and research in the areas of restorative justice, desistance and resettlement potentially enrich the ways in which we theorise, practice and, more generally, seek to encourage offender rehabilitation.

8
The Futures of Rehabilitation

A world without rehabilitation?

In the last two chapters we have tried to identify some promising approaches to rehabilitative work with offenders. Much of this is tentative, and many initiatives have not worked as well as their proponents hoped, for a variety of reasons, but usually not because the fundamental ideas were unsound. The very mixed results of the Crime Reduction Programme (Hough, 2004; Homel *et al.*, 2005) may feed a resurgence of scepticism and a new wave of 'nothing works': the highly sceptical tone of a recent Home Office research review may point in this direction (Harper and Chitty, 2004). To some commentators this might not come as a surprise: they have already warned us that we live in 'dark times' (Nellis, 2004); that a combination of public anxiety about crime and loss of confidence in the capacity of governments to control it can feed an escalation of populist 'toughness' (Garland, 2001); that the growing private sector in corrections, aided by political enthusiasm for 'contestability' and privatisation, has a commercial interest in talking up the problems and escalating the level of coercion (Christie, 1993).

At a political level the hypothesis of 'dark times' can no doubt appear persuasive. As these words are being typed, the Home Secretary is seeking powers to order indefinite house arrest of 'terrorist suspects' without trial or disclosure of evidence, since judges have objected to his earlier practice of indefinite detention of such people without trial in high security prisons. Western countries which were the pioneers of due process and defendants' rights routinely use arbitrary detention, sometimes in secret locations, and inflict humiliating and degrading treatment and interrogation methods widely regarded as torture, in the course of a 'War on Terror' which confronts real threats but has no clearly defined enemy, and no obvious end point. This is apparently a 'war' of the kind that justifies exceptional interference with civil rights, but not of the kind that allows to detainees the rights and protections afforded by international law to prisoners of 'war'. Dark times indeed: such developments hardly create a conducive climate in which to advance arguments for rehabilitation. Indeed, Feeley and Simon have been predicting for some time that rehabilitation will disappear, to be replaced by surveillance and graduated degrees of coercion based on risk assessments (Feeley and Simon, 1992). Whilst they may not offer a fully realistic account of current penal practices (for the reasons discussed in Chapter 6), nevertheless their predictions for the future are certainly gloomy.

However, such arguments tend to present an unbalanced picture. To accept them as a sufficient account would be an error of perspective, caused probably by standing too close to recent developments in Britain. Certainly there have been marked oscillations in official attitudes to rehabilitation; Garland (1995) would argue that this is to be expected, along with regular doses of short-termism and populism. On the other hand, the development of evidence-based rehabilitation has been a long process driven by practitioners and researchers with or without political interest and support: the political dimension can influence the funding and scale of initiatives, and the legislation into which they fit, but it does not determine the development of practice and research. The changes we have reviewed over two centuries show how some of the most promising developments have emerged in the complete absence of official interest: the ideas which exercise a political hegemony at any particular time are not necessarily the same as those which inform the actions of practitioners.

In short, there is no reason for this chapter to add to the burden of penological pessimism. Indeed, there are counter-examples to be found at an international level: for example, most Eastern European countries are abandoning Soviet-style high levels of imprisonment as part of a deliberate policy of moving towards European norms which are, in most cases, less punitive than British practice (and much less punitive than prevailing trends in the USA). However, most of the discussion in this book has been based on developments and evidence from Britain, and this final chapter considers some of the opportunities and arguments which may emerge within the developing penal policy context of Britain.

In England and Wales (though not in Scotland) the future of rehabilitation is closely bound up with the new National Offender Management Service (NOMS). The prison and probation services have been brought

together into a single Service in order to promote 'end-to-end' management of sentences and to facilitate 'contestability' or the involvement of
the private and voluntary sectors in the provision of services (Carter, 2003;
Home Office, 2004a). Introduced suddenly and with minimal consultation, the changes seem to draw on the same political perceptions which
lay behind earlier proposals to amalgamate the two services, and many
commentators are predicting the virtual elimination of the distinctive
identity and traditions of the probation service with its historical
commitment to rehabilitation and a humanistic approach to offenders
(Nellis, 2004). However, although probation services have arguably been
the main carriers of these traditions in recent decades, they have never
been the only carriers, and it would be wrong to overlook the influence
of rehabilitative thinking on the plans for NOMS.

A large part of NOMS' business will be the delivery of the new generic
community sentence introduced by the 2003 Criminal Justice Act, with
its thirteen different kinds of requirement available for the courts to
pick-and-mix.[1] However, the rationale for NOMS is best understood
in relation to another new sentence known as 'custody plus' which
will combine a short prison sentence with a period of supervision in the
community.[2] Currently over 60,000 offenders per year receive custodial
sentences of up to one year in length, and the scandalous neglect of
their needs both inside and outside prison has been one of the significant
drivers behind the NOMS reforms. It is becoming clear that one of the
most promising areas for the future of rehabilitation in Britain is to be
found in the resettlement of short-term prisoners. Some of the issues
and ideas emerging from this were discussed in Chapter 7. In the current
chapter we review some of the implications of recent resettlement
research for 'what works', or 'what should work' in the emerging context
of NOMS.

Rehabilitation, resettlement and the short-term prisoner

The importance attached by the designers of NOMS to resettlement,
and to hybrid sentences of short custodial periods followed by longer
supervision in the community, reflects the relative neglect of this part
of the penal system in the past. The voluntary aftercare of adult prisoners
had its origins in the early nineteenth century, in the work of Discharged
Prisoners' Aid Societies, and did not become a probation service responsibility until this was recommended by the Advisory Council on the
Treatment of Offenders in 1963 (Home Office, 1963). This report also
promoted the term 'throughcare' as an alternative to 'after-care'; the

modern term 'resettlement' was introduced by another Home office report in 1998 (Home Office, 1998). Statutory post-release supervision, with sanctions for non-compliance, has a rather different history, originating in arrangements for the supervision of young offenders released on licence from Borstal institutions (see Chapter 4) and extending to adult prisoners on the introduction of parole in 1967.

The 1991 Criminal Justice Act significantly extended provisions for supervising adult offenders on licence after their release from prison, and in particular introduced Automatic Conditional Release for prisoners serving from one to under four years (Maguire and Raynor, 1997). However, the largest and neediest group of adult offenders, those serving less than one year, were not included in the new provisions, in spite of being also the group most likely to re-offend (Maguire *et al.*, 2000). Historically they had been able to make use of 'voluntary after-care' provided by probation services, and this formed a substantial part of probation officers' work during the 1970s. However, the publication (for the first time) of national priorities for the probation service in 1984 (Home Office, 1984) directed services to put their major effort into statutory supervision of court orders such as probation and community service, in an effort to reduce prison numbers.

The priority accorded to post-custodial supervision consequently fell, and although after 1991 new resources were provided to support the new forms of statutory after-care, by the mid-1990s research showed that voluntary after-care of the short-term prisoner had virtually disappeared (Maguire *et al.*, 2000). When asked by researchers what priority it had, one probation manager replied 'low – should be even lower'. Even among those who were actually offered a service, examination of 53 case files showed only eight who were offered a post-release appointment and kept it. Although some offenders who were interviewed clearly appreciated the service and found it helpful, the majority were missing out and clearly, overall, little was being done.

By the time the Crime Reduction Programme was launched in 1999 this problem was widely recognised, and later reports by NACRO (2000) and the Social Exclusion Unit (2002) were to express similar concerns. The 'pathfinder' programmes set up under the Crime Reduction Programme and described in Chapter 6 included a 'resettlement' pathfinder directed specifically at the short-term prisoner. Consisting of seven projects in all, this aimed to pilot and evaluate services designed to assist the resettlement of short-term prisoners. In essence the intention was to improve the availability and take-up of appropriate post-release services for short-term prisoners, and to help them to make a better transition back into the community on release, with eventual reductions in their future offending.

One other important feature of the resettlement 'pathfinder' was that it was deliberately designed to test and compare a number of different approaches to providing resettlement services. Three voluntary sector projects were to concentrate on 'welfare needs' while four probation-led projects would concentrate more on addressing offending behaviour. For example, by the end of the project all four of the probation-led projects were using a group programme as part of their pre-release provision, and in three cases this was a short cognitive-behavioural programme designed specifically to improve motivation and planning for release.

In a previous discussion of this project (Raynor, 2004b) one of the authors has pointed out that different models of how to provide a resettlement service carry embedded within them different assumptions about why people offend, why they stop offending and how the service provided is intended to influence offending. These ideas are not necessarily made explicit by practitioners, but they constitute in effect an implicit criminology or a set of implicit criminologies corresponding to each model of service provision. The most obvious contrast is between what might be described as an 'opportunity deficit' model and an 'offender responsibility' model. Table 8.1 illustrates the differences.

Both these models are ideal types which are not necessarily found in a pure form in the real world, but they do bear some relation to the different approaches and assumptions actually encountered when studying resettlement practice. The first (the opportunity-deficit model) sees offenders as offending because they have been deprived of resources

Table 8.1 Contrasting models of resettlement

	Opportunity deficit model	Offender responsibility model
Methods	Advice, support, sympathy, advocacy, referral	As left-hand column plus cognitive challenge, motivational and pro-social input
Assumptions about offending	Offenders are victims of circumstances	Offenders can choose how they respond to circumstances
Offender's responsibility for offending and change	Low	High

Source: Raynor (2004b).

and opportunities, and concentrates on putting them into contact with agencies which it is hoped will supply these. This type of service basically responds to offenders' own statements about their needs, and reflects the accounts many offenders give of how they got into trouble through lack of some resource or blockage of some opportunity. Such accounts can contain a good deal of truth, but tend to allow offenders to present themselves as victims of circumstances who had little alternative but to offend. This can also be an aspect of their 'self-talk' or the way they describe their offending and their lives to themselves. Sykes and Matza nearly fifty years ago identified this type of rationale as a 'neutralisation technique' (Sykes and Matza, 1957): they discuss a number of variants of this type of process, which all have the effect of making offending more acceptable to the offender.

The alternative 'offender's responsibility' approach recognises that social, environmental and personal problems are real, but tends to treat them as challenges or obstacles which confront offenders with choices about how to respond. This is the model implied by phrases such as 'confronting offending' and 'challenging offenders' thinking'. It tends to place an emphasis on helping offenders to develop clear goals for the future, and to acquire the problem-solving skills, self-efficacy and motivation to overcome some of the obstacles they are likely to face. Such approaches may sometimes run the risk of underestimating the extreme difficulties many offenders encounter, but they have the merit that they present further offending as avoidable, and offenders as capable of desisting.

Resettlement and desistance

These contrasting approaches bear some similarity to the differences built into the pathfinder projects between those which were intended to concentrate on 'welfare' problems, and those which were intended to take a more proactive approach to challenging offending behaviour and the associated thinking. Pending the availability of a reconviction study, the seven projects were evaluated using two main measures: changes in scores on the CRIME-Pics II instrument (Frude *et al.*, 1994: designed to measure anti-social and crime-prone attitudes and beliefs) and continuity (the extent to which prisoners remained in contact with helping services after release). Readers interested in the detail of the findings will find them in the full report (Lewis *et al.*, 2003), but in essence both measures showed generally better results in the probation-led projects. One possible interpretation of these results (others are of course possible) is that the most promising approaches to resettlement have more in common with the 'responsibility' model outlined above than with the 'opportunity deficit' model.

Two recent criminological accounts of the processes of recidivism and desistance tend to support this view. The first concerns what its authors call the 'criminal recidivism process' (Zamble and Quinsey, 1997) and the second is based on offenders' own accounts of recidivism and desistance (Maruna, 2000). Both these studies point, in different ways, to the fact that future offending is likely to be influenced by offenders' thinking as well as their circumstances. Zamble and Quinsey's study of released male prisoners in Ontario found that the process of recidivism often involved practical or situational difficulties leading to negative emotion (depression, anger) and perhaps pessimism about the possibility of dealing with the situation other than by offending. This led people to give up on their attempts to avoid offending. As Zamble and Quinsey put it, 'in the case of criminal behaviour, factors in the social environment seem influential determinants of initial delinquency for a substantial proportion of offenders...but habitual offending is better predicted by looking at an individual's acquired ways of reacting to common situations' (pp. 146–47). This suggests that services for released prisoners must address not only the problems they face but also the strategies, motivation, personal resources and skills they will need to deploy in overcoming them.

Maruna's (2000) interview-based study of offenders in Liverpool, which we discussed in Chapter 7, suggested that these personal resources are related to the way offenders themselves understand and account for their situation and behaviour. He describes these understandings and accounts as different kinds of narrative, some of which support continued offending and some of which support desistance. For example, those who were continuing to offend saw themselves as victims of circumstances who had little choice, while those who were desisting from offending saw themselves as having taken control of their lives and determining their own futures. In discussing the latter group, Maruna describes the 'narrator's strong sense that he or she is in control of his or her destiny. Whereas active offenders in the Liverpool Desistance Study...seemed to have little vision of what the future might hold, desisting interviewees had a plan and were optimistic that they could make it work' (p. 147).

Studies such as this lend some support to the idea that rehabilitative services should address both opportunities and thinking: in other words, they should aim to reinforce and support plausible 'narratives of desistance'. There may even be dangers in an exclusive focus on problems of access to resources and opportunities, as if crime were *nothing more than* a response to environmental difficulties: this may run the risk of reinforcing recidivist 'narratives' in which offenders cast themselves as victims of circumstance. The providers of rehabilitative services may have a choice between reinforcing narratives which tend to support desistance, and reinforcing narratives which tend to support or facilitate re-offending. Clearly the former is preferable. The provision of resettlement services also requires efficient links between what happens inside and outside prison, and between 'offender management' services and the many other agencies in the community which released prisoners will need to use in addressing problems of accommodation, education, employment or substance abuse. The NOMS approach to integrated services might offer some real advantages, at least to those offenders who currently undergo short prison sentences.

As the previous chapter noted, Maguire (2004b) has welcomed the emergence of regional resettlement schemes which bring all the relevant agencies together in a joint regional strategy. Although the NOMS proposals have been subject to many criticisms and clearly represent a threat to the future of probation services, and they also provide a good example of what Cohen (1985) described as a tendency to blur the boundaries between custodial and non-custodial sanctions, they could also represent a serious opportunity to improve the rehabilitative services available to released prisoners and even to increase the rehabilitative efforts made within prisons themselves. In fact, a number of experienced commentators have argued that the prospects for a better co-ordinated and more rehabilitative system are improved by NOMS (for example, Faulkner, 2004), and much will depend on how the new proposals are implemented in practice.

Strengths-based models and community involvement

The previous chapter referred to the emergence in America and (to a so far lesser degree) in Britain of 'strengths-based' models (Maruna and LeBel, 2003). These have something in common with the 'responsibility approach' outlined above. Such approaches see offenders not simply as objects of 'treatment' or 'intervention', characterised by needs and deficits and presenting risks, but as active participants in their own rehabilitation, with strengths and skills and potential as contributors to their communities. This is connected with that meaning of 'rehabilitation', discussed in Chapter 1, which implies restoration of moral status as a full citizen and a contributing member of society. There is also an evident overlap with the concept of reintegration, and with the ideas for constructive handling of crime advanced particularly in America under the label of 'community justice' (Clear and Karp, 1999). Treating offenders as needy and beset by problems perhaps corresponds too closely with some of their own rationales and justifications for offending (Sykes and Matza, 1957), whilst treating them as able to take charge of their lives, overcome obstacles and help others could promote a sense of responsibility and self-efficacy.

Some approaches to rehabilitation have explicitly treated ex-offenders as experts, and as people able to put their experience to use in helping others. The New Careers movements of the 1970s and 1980s aimed to retrain former offenders for social work roles, often with offenders (Seddon, 1979), and the involvement of recovered substance misusers in helping others to follow the same path has been a central feature of self-help movements such as Alcoholics Anonymous, where it is seen as part of recovery and of remaining sober. A more recent example is the Lifelines project in Canada, in which the resettlement of released lifers in the community after long periods of imprisonment is assisted by successfully resettled former lifers who act as mentors, supporters and advocates (www.csc-scc.gc.ca/text/prgm/lifeline). This is reminiscent of Braithwaite's argument (Braithwaite, 2001) that people generally expect to see offenders taking 'active responsibility' in making positive contributions to the community. Similarly, Hans Toch has recently argued that involvement in altruistic activity itself promotes cognitive change and pro-social learning (Toch, 2000), pointing to yet another overlap between the restorative and the rehabilitative agendas.

Two points of clarification are required here. First, the 'responsibility' which strengths-based models give to offenders or expect from them is to be understood as the normal human freedom to make choices and to be accountable for them, even in circumstances which are not of one's own choosing. The degree of the offender's responsibility for offending may depend partly on what other options and possibilities were available, and what opportunities existed to avoid offending; but the process of choosing and actively constructing an offence free or pro-social lifestyle requires choices from the offender, and often great determination. This is not the same as the notion of 'responsibilization' which is identified by Rose (2000) and others as part of the strategy of neo-liberal restructuring of the role of the State, based on economic individualism and the withdrawal of State responsibility for welfare. The aim of such ideological strategies is to force citizens to take responsibility for their own welfare in a free market: in other words, to transfer responsibility from government to the individual, whether or not individuals are actually facing problems which can realistically be addressed at an individual level. Some writers (for example, Kendall, 2004) have tended to see this kind of responsibilization at work in modern approaches to rehabilitation; however, in our view the development of strengths-based approaches requires the active involvement of the State and government agencies in promoting and resourcing the policies and practices which might make this a reality.

In addition, to reiterate a point we made in Chapter 7, we need to be cautious about the word 'community', often misused to cast a warm glow over what are actually poorly thought out and poorly resourced policies (Raynor, 2001). Some policies, such as 'Care in the Community', actually rely for their persuasiveness on a fatal and perhaps deliberate confusion between 'community' meaning anywhere outside institutions, and 'community' meaning locality-based social systems of neighbourliness, reciprocity and mutual support. Many localities are 'communities' in the first sense but not the second; others may contain distinct 'communities' which are mutually hostile or actually in conflict, or dominant 'communities' committed to the subjection of others. Expecting the 'community' to contribute to the management of social problems does not avoid the need to think about and promote or strengthen the methods, institutions and social practices which might make this possible.

The previous chapter showed how rehabilitation can be understood and developed as a relational process rooted in communities: initiatives such as the re-entry court offer a form of 'therapeutic jurisprudence' (McGuire, 2003) which engages magistrates, as representatives of the community, in a process of continuing dialogue with the offender as part of the rehabilitation process. Some similar processes have been seen at work in drug courts where sentencers maintain an interest in and a commitment to offenders and their progress, and become involved in a problem-solving approach to difficulties and lapses (McIvor *et al.*, 2003). Other vehicles for community involvement range from 'circles of support and accountability' assisting the reintegration of sex offenders who are trying to avoid re-offending (Correctional Service of Canada, 2003) to involvement of neighbourhood panels or of victims of crime in selecting reparative work to be done by offenders as part of a community sentence.[3]

In a small-scale but particularly interesting example, the Channel Island of Jersey combines an evidence-driven 'what works' approach to community penalties (Heath *et al.*, 2002) with a centuries-old system of community crime control by elected parish officials (Miles, 2004) in which reparation and reintegration are dominant themes. The island's Probation and After-Care Service is actively involved in both approaches, which are seen by practitioners as mutually supportive and informed by the same principles. This is just one example of how effective practice is generated and sustained at a local level, through face-to-face interactions with offenders and community members (Faulkner and Flaxington, 2004). The essential work of rehabilitation is done in localities rather than in Parliament or Downing Street, or even in the Home Office.

Responsibility and the State

Whilst we have little doubt that such ideas will attract growing attention among academics and practitioners and will continue to inform many projects and experiments, the extent to which they inform large-scale policies and developments, and therefore the extent to which they affect the opportunities and chances of large numbers of offenders, depends inescapably on the politics of criminal justice, and on the politicians. We are often concerned about how to 'sell' rehabilitation to politicians and the public, and to some extent this reflects concerns about its perceived legitimacy: it is not easy to sell the idea of offenders receiving benefits and services which are denied to similarly needy people who have not offended. This, however, is an image of rehabilitation which flows from the deficit model. If instead we see rehabilitation as work undertaken by offenders to improve or restore their own capacity for pro-social involvement in the community, this is a contribution to community well-being, and we have argued that one function of rehabilitative services is to offer opportunities to do this kind of work. Our interest in theories of state-obligated rehabilitation derives partly from the rationale they offer for this kind of approach.

The idea of state-obligated rehabilitation as expounded by Carlen (1989) or Hudson (2003, Chapter 4) appears to rest on a version of social contract theory: the State's demand that citizens should obey the law derives part of its moral legitimacy from the extent to which the State accepts an obligation to prevent extremes of poverty or deprivation, and to promote opportunities and resources which make crime-free living a realistic and achievable aspiration. Social contract theory suggests that if offenders have had limited opportunities to lead crime-free lives or to learn how this is done, action by the State to improve these opportunities puts them on a more equal basis with other citizens rather than a privileged basis, provided that they are prepared to reciprocate through a serious attempt at self-rehabilitation. This suggests that rehabilitative policies will be at their most effective when underpinned by a commitment to universalist welfare policies which offer assistance to all citizens who need it, *including* (in every sense) offenders. It also suggests, in line with the arguments set out in the previous chapter, that rehabilitation is best understood not primarily as the prevention of re-offending, but as the promotion of desistance from offending. Rehabilitation then becomes something done *by* the offender rather than *to* the offender.

The moral discourse of rehabilitation

At the beginning of this chapter we argued that fears about the survival of rehabilitation are probably misplaced. Our survey in previous chapters of a range of rehabilitative ideas and practices suggests a number of utilitarian reasons why it should survive, and ways in which it can benefit both individuals and communities. However, as we discussed in Chapter 2, arguments about rehabilitation are not simply about instrumental utility. They also make implicit or explicit claims about the nature and value of persons, about the relationship between states and citizens, and about human rights.

Some years ago, journals with an interest in probation and criminal justice regularly carried articles about the probation service's values (for example, James, 1995; Nellis, 1995; Spencer, 1995). In one particularly interesting discussion of how probation activities might be evaluated, McWilliams (1989) suggested that some organisations and institutions are important not simply for what they achieve, but for the values they express and embody. The probation service, he argued, might be seen as such an expressive organisation and evaluated accordingly. Similarly we would argue that rehabilitative services and procedures represent particular values within the criminal justice system, without which the system would be morally impoverished as well as less effective. Belief in the value and rights of all persons, and in their potential for change, is an important safeguard against the dominance of punitive or 'eliminative' approaches (Rutherford, 1998) and forms an important component in the 'decency agenda' (Rutherford, 1994) which prison reformers have tried to implement particularly since the Woolf Report (1991).

Rehabilitation requires us to get to know individual offenders: their needs, their weaknesses, their strengths, their aspirations, how the world looks and feels to them. This makes it psychologically harder for most people to treat them with indifference or to inflict cruelty and degradation. In the absence of a commitment to rehabilitation, punishment can easily tip over into inhuman or degrading treatment. Recently one newspaper reported allegations that prison officers in a young offenders' institution had been playing a game called 'Gladiator', which consisted of allocating to shared cells prisoners who were likely to fight each other (for example, an Asian prisoner and a white racist) and then betting on the outcome (*Guardian*, 4th March 2005). If a commitment to rehabilitation helps to avoid this kind of behaviour, that is some justification in itself.

Garland has suggested that modern criminologies can be divided into criminologies of the 'self', which emphasize how criminals resemble the rest of us ('normal, rational consumers, just like us', Garland, 2001, p. 137) and criminologies of the 'other' which portray criminals as 'the threatening outcast, the fearsome stranger, excluded and embittered'. He looks for a middle ground between these two poles, which is represented by an older 'social democratic' and welfarist criminology. (He also points out that this earlier criminology has not been scientifically discredited: perhaps some of the evidence we have reviewed lends it some scientific support.) Another view might be that this middle ground represents a more realistic 'criminology of the self': are we always normal and rational, never threatening or embittered or in other ways anti-social? The social democratic tradition implies an awareness of social context, and what might be called a criminology of the potential self, or the unlucky self: in other words, the selves we might have been if brought up in less favourable circumstances or with fewer advantages. The social solidarity represented by the social democratic tradition is based on awareness of shared humanity and a shared human predicament.

Something like this was vividly expressed more than eighty years ago by the Lebanese writer Kahlil Gibran, in a poem about 'crime and punishment' (Gibran, 1991, pp. 52–55):

Oftentimes have I heard you speak of one who commits a wrong as though he were not one of you, but a stranger unto you and an intruder upon your world.

But I say that even as the holy and righteous cannot rise beyond the highest which is in each one of you,

So the wicked and the weak cannot fall lower than the lowest which is in you also...

...and when one of you falls down he falls for those behind him, a caution against the stumbling stone...

And he falls for those ahead of him, who, though faster and surer of foot, yet removed not the stumbling stone.

In more prosaic language, justice, atonement and reintegration draw on a sense of shared humanity and of a connection between the self and the other. Hudson (2003) discusses the problem of why we should respect the rights of those unlike ourselves: what secures decent treatment for those with whom we cannot easily identify, and for whom we may feel little sympathy or respect? She quotes the view of the philosopher Richard Rorty that what is needed is a 'progress of sentiment' which consists in 'an increasing ability to see the similarities between others and ourselves as outweighing the differences. Rorty stresses that the similarities he has in mind are not the abstractions of an essential human nature, but the everyday similarities of caring for our parents or children, and grieving when they are hurt' (Rorty, 1993, p. 129; Hudson, 2003, p. 223). Such awareness of similarity develops most easily when there is contact and communication.

All of these examples are ways of trying to express the values of human solidarity which inform not only the rehabilitative ideal, but also a humane approach to criminal justice, and indeed the possibility of organised and peaceful societies. At its best, rehabilitation aspires to make these values real and effective in the social practices of the criminal justice system. We now know more about how to help offenders effectively than we did at any time in the past, and we have learned a good deal about the difficulties of making it happen in practice, and a little about how to overcome them. Twenty years ago, hardly anyone would have predicted that such achievements were imminent. The practices associated with rehabilitation change over time, but the ideals are remarkably durable. Whatever structural changes are made in the correctional system, we hope and expect to see a continuing commitment to rehabilitation in the coming decades.

Notes

1 Defining rehabilitation

1. It should be noted, however, that the 'resettlement' process is often thought to include work done with the offender (and possibly his or her family) prior to release, such that the separation of resettlement and punishment is not always or necessarily clear-cut (e.g. Home Office, 2001, p. 12).

3 Origins and contexts

1. In 1776 Howard estimated that over half of all prisoners in England and Wales were debtors.
2. Namely, those guilty of murder, setting houses on fire, and house-breaking 'attended with acts of cruelty' (Howard, 1929, p. 261).
3. Fielding argued that 'moral correction' would only be achieved through the influence of religion, and that this was a matter 'of the highest Consequence' (1753, p. 258).
4. Beccaria summed up his thesis as follows: 'In order for punishment not to be, in every instance, an act of violence of one or of many against a private citizen, it must be essentially public, prompt, necessary, the least possible in the given circumstances, proportionate to the crimes, dictated by the laws' (Beccaria, 1963, p. 99).
5. Bentham proposed that the majority of the profits of prisoners' labour would be devoted to maintaining the institution itself, thereby sparing the public purse, whilst one quarter of the profits should go to prisoners, a proportion of which should be converted into annuities for their future benefit (McConville, 1981, p. 119).
6. Ignatieff notes that in drafting the Act the term 'penitentiary' came to replace the term 'hard labour house', and that this substitution 'seemed to express the ideal of a place not merely of industry but also of contrition and penance' (1978, p. 94).
7. It is interesting to note that Becher opposed the idea of a national penitentiary not only on financial grounds but also because he thought it undesirable to place prisoners in an institution remote from their own neighbourhoods (McConville, 1981, p. 132).
8. The construction of Pentonville was part of the transition from transportation to a national penal system for long-term prisoners which occurred between the 1840s and the 1850s. See Ignatieff, 1978, Chapter 1, for a detailed account of Pentonville's regime.
9. Darwin's *Origin of the Species* was published in 1859, and reportedly played a major role in brining evolutionary theory into the mainstream of British intellectual thought (Forsythe, 1991; Vanstone, 2004).

10. After 1848 the government began to replace transportation with a national system of convict prisons. By 1863 imprisonment had been transformed into a punishment for all the major crimes, except murder, which were formerly punished by either hanging or transportation. Under the new system of 'penal servitude', convicts were sent to work at one of a number of 'public-works prisons'. This period of hard labour followed a six-month period of solitary confinement at Pentonville or Millbank. At the end of their sentence, convicts were released on a 'ticket of leave', a system which is commonly understood as the forerunner of the modern system of parole (Shute, 2003). This required them to report to the police at regular intervals, maintain a steady job and avoid association with other offenders (Ignatieff, 1978, p. 201).

11. The Prison Commission was set up to take over the running of the local prisons. By the early 1890s Du Cane also chaired the Directorate of convict prisons which had been set up in 1850 following the cessation of transportation.

12. As Radzinowicz and Hood observe, Du Cane's position was that of Continental classicism, although they doubt that Du Cane would have been aware of this ideological affinity.

13. It was intended to serve two purposes: adding the deterrent element that transportation was thought to lack and providing occupational training to enable fit young convicts to make a better life for themselves in the colonies. Although bound for transportation, assessments of prisoners' behaviour determined whether the individual was to be transported as a free emigrant, or under a conditional pardon, or, if deemed incorrigible, he was to be kept in confinement on arrival (McConville, 1981).

14. According to Radzinowicz and Hood, Vincent's proposal that probationers be placed under the supervision of the police 'led to bitter controversy, bred confusion, encouraged ambivalence, and without doubt played an important part in delaying the introduction of probation for two decades' (1990, pp. 636–37).

15. Whilst clearly prompted by a scathing attack on Du Cane's regime in the national press (see p. 40), the establishment of the Departmental Committee under Herbert Gladstone has to be understood in the context of a particular political mood; a 'new Liberalism' which was at that time raising questions about the administration of a number of different institutions (Radzinowicz and Hood, 1990, p. 575; McConville, 1995b, p. 549). This political climate meant that the time was right for a review of the penal system which would go quite some way beyond its rather narrow terms of reference, essentially questioning not just the effectiveness but also the underlying objectives of the prevailing system of punishment (Allen, 1995).

16. McConville (1995a, p. 157) argues that Gladstone gave this proposal 'a quite insupportable prominence' and that the recommendation masked some quite different opinions about the effectiveness of existing reformatories among the witnesses interviewed by the Committee.

4 The rehabilitative ideal: Advance and temporary retreat

1. One of the earliest seems to have been a course of lectures offered in the Women's University Settlement in Southwark as early as 1892 (Macadam, 1925).

2. 'Systems theories' were advanced by a number of writers on social work but their best-known exponents were Pincus and Minahan (1973) who tried to break down the then traditional distinction between casework, groupwork and community work by seeing them all as a form of planned intervention in smaller or larger social systems. In particular, they distinguished between the 'client system' (the beneficiary), the 'target system' (the object of intervention), the 'action system' (the resources used to effect change) and the 'change agent' (usually the social worker). The idea of a single 'unitary' social work method with a range of applications excited the Central Council for Education and Training in Social Work, and the approach offered an interesting framework for assessment and analysis, but it offered little in the way of testable methods and fairly quickly went out of fashion. It was, however, one of the influences which contributed to the useful and effective 'system approach' used by juvenile justice practitioners in Britain in the 1980s to reduce custodial sentencing and residential care of young offenders (see Chapter 5).

5 Adapting to the end of 'treatment'

1. Bottoms explains that the term 'populist punitiveness' is not the same thing as 'public opinion': rather, it is 'intended to convey the notion of politicians tapping into, and using for their own purposes, what they believe to be the public's generally punitive stance' (1995, p. 40).
2. The case of Christopher Clunis, a discharged psychiatric patient who, in 1992, stabbed Jonathan Zito to death in a London underground station, attracted considerable publicity.
3. In the 1988 Action Plan 'success' was defined as reducing the use of custody with juveniles, not necessarily impacting on their behaviour.
4. Whilst reform in this context appears outdated – a throwback to earlier times (see Chapter 3) – it can also be understood as an attempt to create a distance between contemporary practices and the 'treatment' approaches commonly associated with the term 'rehabilitation'. Thus, where 'reform' is discussed in the White Paper, it is with reference to 'non-treatment' concepts such as self-control, responsibility, victim awareness and citizenship. Reparation, in the form of community service, is also portrayed in the White Paper as a potentially reformative sanction.

6 The new rehabilitation: 'What Works' and corrections at the end of the twentieth century

1. 'Failure' meant a premature end to a placement due to absconding or re-offending. Absconding and reoffending were also strongly correlated with each other.
2. One subgroup which did *better* under intensive supervision consisted of probationers with a high level of self-reported problems but a low level of 'criminal tendencies', as assessed by probation officers. This suggests that the methods used by probation officers resembled counselling, and benefited this

particular group: unfortunately they were a small minority, and untypical of offenders as a whole – see Chapter 4.

3. These combined the requirements of a probation order with those of a Community Service Order.

4. The case of suspended sentences was special, in that the court was *required* to sentence custodially on breach except in very narrowly defined exceptional circumstances.

5. For example, the STOP programme in Mid Glamorgan was preceded by a year of planning and genuine consultation with staff about the initiative, and all probation officers were trained to deliver the programme. This meant that although at any given time only some of the trained officers were running programme groups, the remaining officers knew what the programme was for and how it was meant to work (Raynor and Vanstone, 1997).

6. This was a large government initiative which provided new funding from 1999 to 2003 in order to test a wide range of approaches to crime reduction, ranging from crime prevention measures to post-sentence interventions such as the Probation Service's pathfinders. For a full account see Hough 2004 and Homel *et al.*, 2005.

7. Some findings of this study in relation to resettlement and desistance from offending are discussed in more detail in Chapter 8.

8. In any case it is usually very difficult to know this with much accuracy, since highly dangerous offences are relatively rare and this makes their prediction statistically difficult.

9. Women are not, of course, a minority in the general population but they are in the correctional population because of their lower propensity to offend.

7 Against the tide: Non-treatment paradigms for the twenty-first century

1. As Dignan himself notes, rehabilitating offenders and meeting the needs of victims are not mutually exclusive: many victims are likely to be interested in helping to prevent further victimization.

2. The Home Office has subsequently published a 'restorative justice strategy' (Home Office, 2003) and, building on experience in Thames Valley, the Criminal Justice Act 2003 has introduced a statutory basis for restorative justice as part of a police caution. The Home Office strategy document makes it clear that the Government sees restorative justice as an effective way of putting victims' needs at the heart of the criminal justice system, as well as encouraging offenders to take responsibility and make amends for their behaviour, and for involving the community in responses to crime and anti-social behaviour.

3. However, far from being 'soft' on offenders, restorative justice is often an uncomfortable experience for offenders, and requirements in respect of 'making amends' may be burdensome (Johnstone, 2002, p. 93). In Walgrave's words, 'restorative justice is not a soft option' (2001, p. 17).

4. One exception to this is the practice of early termination of a community (probation) order in the light of good progress.

5. Reflecting on the recent 'coming together' of two formerly separate areas of criminological research (one on desistance; the other on effective correctional interventions, or 'what works'), Farrall and Maruna note that desistance researchers have long been aware of the potential relevance of their work to the realm of 'expert' correctional interventions:

> The Gluecks put the question this way: 'can educators, psychologists, correctional workers, and others devise means of "forcing the plant", as it were, so that benign maturation will occur earlier than it seems to at present?' (Farrall and Maruna, 2004, pp. 360–61)

6. Rex (1999) also found that the probationers in her study valued guidance concerning their personal and social problems, an aspect of probation intervention which Rex summarised as 'strengthening social ties'.

7. Indeed, in her foreword to Stephen Farrall's book, Christine Knott (then Chief Probation Officer, Greater Manchester Probation Area) hints at the same tendency (in Farrall, 2002, pp. xiii–xiv).

8. Taking his regressive/progressive argument still further, McNeill (2004) notes that desistance research seems to make a good case for the rehabilitation of probation's original moral purpose: to 'advise, assist and befriend'.

9. At least since a government report proposed this term to replace 'throughcare', possibly because of the unfashionable notion of caring implied by the latter (Home Office, 1998).

10. Another contemporary initiative which fits this 'devolved' model is 'Circles of Support', which relates to the management of sex offenders in the community. Developed in Canada but currently being piloted in the UK, this initiative draws together community members who, on a voluntary basis, offer social support to the sex offender whilst also acting as monitors of potentially risky behaviour. This initiative thereby broaden the responsibility for sex offender rehabilitation (and, importantly, surveillance) beyond the statutory agencies of police and probation services (Petrunik, 2002; Kemshall and Maguire, 2003).

8 The futures of rehabilitation

1. These include requirements relating to unpaid work, specified activities, accredited programmes, prohibited activities, curfews, exclusions from specified places, residence in specified places, mental health treatment, drug rehabilitation and alcohol treatment, supervision by a responsible officer, attendance centres for younger offenders, and electronic monitoring.

2. Both the length of the overall sentence and the length of the custodial period will be specified by the court. Any prison sentence up to 51 weeks in length will be served as a mixed sentence, with the custodial part lasting from two weeks (in a 28-week sentence) to 13 weeks (in a 51-week sentence), followed by a period under supervision in the community ranging from

26 weeks in a short sentence to 38 weeks at the longest end of the range (or even more if the court specifies a 51-week sentence with a custodial period of less than 13 weeks).

3. Suggested as a possible way to increase public support for community sentences (Raynor, 2001), this is now official policy in Scotland (Scottish Executive, 2004) as well as in England and Wales, where it was reported by one newspaper as 'Neighbours to decide punishments' (*Guardian*, 8 December 2004).

References

Abel-Smith, B. and Townsend, P. (1965) *The Poor and the Poorest*, London: Bell.

Adams, S. (1961) 'Interaction between individual interview therapy and treatment amenability in older youth authority wards', in *Inquiries Concerning Kinds of Treatment for Kinds of Offenders*, Sacramento: California Board of Corrections, pp. 27–44.

Adams, S. (1967) 'Some findings from correctional caseload research', *Federal Probation*, 31, 4, 48–57.

Advisory Council on the Penal System (1968) *The Regime for Long-term Prisoners in Conditions of Maximum Security* (Radzinowicz Report), London: HMSO.

Advisory Council on the Penal System (1970) *Non-custodial and Semi-custodial Penalties* (Wootton Report), London: HMSO.

Advisory Council on the Treatment of Offenders (1963) *The Organisation of After-Care*, London: HMSO.

Allam, J. (1998) *Effective Practice in Work with Sex Offenders: A Reconviction Study Comparing Treated and Untreated Offenders*, Birmingham: West Midlands Probation Service.

Allen, F.A. (1959) 'Criminal justice, legal values and the rehabilitative ideal', *Journal of Criminal Law, Criminology and Police Science*, 50, 226–32.

Allen, F.A. (1981) *The Decline of the Rehabilitative Ideal: Penal Policy and Social Purpose*, New haven: Yale University Press.

Allen, Lord of Abbeydale (1995) 'In search of the purpose of prison', in Prison Reform Trust, *Gladstone at 100: Essays on the Past and Future of the Penal System*, London: Prison Reform Trust.

American Friends Service Committee (1971) *Struggle for Justice: A Report on Crime and Punishment in America*, New York: Hill and Wang.

Andrews, D.A. (2001) 'Principles of effective correctional programming', in L. Motiuk and R. Serin (eds) *Compendium 2000 on Effective Correctional Programming*, Ottawa: Correctional Service of Canada.

Andrews, D.A. and Bonta, J. (1998) *The Psychology of Criminal Conduct*, Cincinnati: Anderson.

Andrews, D.A., Zinger, I., Hoge, R.D., Bonta, J., Gendreau, P. and Cullen, F.T. (1990) 'Does correctional treatment work? A clinically relevant and psychologically informed meta-analysis', *Criminology*, 28, 369–404.

Association of Chief Officers of Probation (1994) *Guidance on Management of Risk and Public Protection*, London: ACOP.

Attlee, C. (1920) *The Social Worker*, London: Bell.

Ayscough, H.H. (1923) *When Mercy Seasons Justice: A Short History of the Works of the Church of England in the Police Courts*, The Church of England Temperance Society.

Bailey, R. and Brake, M. (eds) (1975) *Radical Social Work*, London: Edward Arnold.

Bale, D. (1987) 'Uses of the risk of custody scale', *Probation Journal*, 34, 4, 127–31.

Bazemore, G. (1996) 'Three paradigms for juvenile justice', in B. Galaway and J. Hudson (eds) *Restorative Justice: International Perspectives*, Monsey, NY: Criminal Justice Press.

Bazemore, G. (1998) 'Restorative justice and earned redemption', *American Behavioral Scientist*, 41, 6, 768–813.

Bazemore, G. (1999) 'After shaming, whither reintegration: Restorative justice and relational rehabilitation', in G. Bazemore and L. Walgrave (eds) *Restorative Juvenile Justice: Repairing the Harm of Youth Crime*, Monsey, NY: Criminal Justice Press.

Bazemore, G. and Bell, D. (2004) 'What is the appropriate relationship between restorative justice and treatment?', in H. Zehr and B. Toews (eds) *Critical Issues in Restorative Justice*, Monsey, NY: Criminal Justice Press.

Beccaria, C. (1963) [1764] *On Crimes and Punishment* (trans. H. Paolucci), Indianapolis: Bobbs-Merrill.

Beck, U. (1992) *Risk Society: Towards a New Modernity*, London: Sage.

Becker, H. (1963) *Outsiders*, New York: Free Press.

Becker, H. (ed.) (1973) 'Labelling theory reconsidered', *Outsiders* (revised edition), New York: Free Press.

Becket, R., Beech, A., Fisher, D. and Fordham, A. (1994) *Community-based Treatment For Sex Offenders: An Evaluation of Seven Treatment Programmes*, London: Home Office.

Beech, A., Fisher, D. and Becket, R. (1999) *STEP 3: An Evaluation of the Prison Sex Offender Treatment Programme*, London: HMSO.

Behan, B. (1958) *Borstal Boy*, London: Hutchinson.

Bentham, J. (1791) 'Panopticon; or, The Inspection House', in P. Rock (ed.) (1994) *History of Criminology*, Aldershot: Dartmouth.

Bentham, J. (1823) 'Principles of Morals and Legislation', reprinted in W. Harrison (ed.) *A Fragment on Government with an Introduction to the Principles of Morals and Legislation*, Oxford: Blackwell (1948).

Bernfeld, G., Farrington, P. and Leschied, A. (eds) (2001) *Offender Rehabilitation in Practice*, Chichester: Wiley.

Berntsen, K. and Christiansen, K. (1965) 'A resocialization experiment with short-term offenders', *Scandinavian Studies in Criminology*, 1, 35–54.

Biestek, F. (1961) *The Casework Relationship*, London: Allen and Unwin.

Blackburn, R. (1980) 'Still not working? A look at recent outcomes in offender rehabilitation', Paper presented at the Scottish Branch of the British Psychological Society Conference on Deviance, University of Stirling.

Blumsom, M. (2004) 'First steps and beyond: The pathway to our knowledge of delivering programmes', *VISTA*, 8, 171–76.

Bochel, D. (1976) *Probation and After-Care: Its Development in England & Wales*, Edinburgh: Scottish Academic Press.

Bonta, J. (1996) 'Risk-needs assessment and treatment', in A. Harland (ed.) *Choosing Correctional Options that Work*, London: Sage.

Booth, W. (1890) *In Darkest England and the Way Out*, London: The Salvation Army.

Blom-Cooper, L. and McConville, S. (1995) 'In the light of modern knowledge...', in Prison Reform Trust, *Gladstone at 100: Essays on the Past and Future of the Prison System*, London: Prison Reform Trust.

Bottomley, A.K. (1980) 'The "justice model" in America and Britain: Development and analysis', in A.E. Bottoms and R.H. Preston (eds) *The Coming Penal Crisis*, Edinburgh: Scottish Academic Press.

Bottoms, A.E. (1977) 'Reflections on the renaissance of dangerousness', *The Howard Journal*, 16, 2, 70–96.

Bottoms, A.E. (1980a) 'An introduction to "the coming crisis"', in A.E. Bottoms and R.H. Preston (eds) *The Coming Penal Crisis*, Edinburgh: Scottish Academic Press.

Bottoms, A.E. (1980b) *The Suspended Sentence after Ten Years*, Leeds: Centre for Social Work and Applied Social Studies, University of Leeds.

Bottoms, A.E. (1983) 'Neglected features of contemporary penal systems', in D. Garland and P. Young (eds) *The Power To Punish: Contemporary Penality and Social Analysis*, London: Heinemann Educational Books.

Bottoms, A.E. (1995) 'The philosophy and politics of punishment and sentencing', in C. Clarkson and R. Morgan (eds) *The Politics of Sentencing Reform*, Oxford: Clarendon Press.

Bottoms, A.E. and McWilliams, W. (1979) 'A non-treatment paradigm for probation practice', *British Journal of Social Work*, 9, 159–202.

Bottoms, A.E., Brown, P., McWilliams, B.C., McWilliams, W.W. and Nellis, M. (1990) *Intermediate Treatment and Juvenile Justice*, London: HMSO.

Braithwaite, J. (1989) *Crime, Shame and Reintegration*, Cambridge: Cambridge University Press.

Braithwaite, J. (1993) 'Shame and modernity', *British Journal of Criminology*, 33, 1, 1–18.

Braithwaite, J. (1999) 'Restorative justice: Assessing optimistic and pessimistic accounts', in M. Tonry (ed.) *Crime and Justice: A Review of Research*, 25, Chicago: University of Chicago Press.

Braithwaite, J. (2001) 'Intention versus reactive fault', in N. Naffine, R. Owens and J. Williams (eds) *Intention in Law and Philosophy*, Aldershot: Ashgate.

Braithwaite, J. and Mugford, S. (1994) 'Conditions of successful reintegration ceremonies', *British Journal of Criminology*, 34, 2, 139–71.

Brake, M. and Bailey, R. (eds) (1980) *Radical Social Work and Practice*, London: Edward Arnold.

Brewer, C. and Lait, J. (1980) *Can Social Work Survive?*, London: Temple Smith.

Brody, S.R. (1976) *The Effectiveness of Sentencing*, Home Office Research Study, 35, London: HMSO.

Bryant, M., Coker, J., Estlea, B., Himmel, S. and Knapp, T. (1978) 'Sentenced to social work', *Probation Journal*, 25, 110–14.

Burnett, R. (1992) *The Dynamics of Recidivism*, Oxford: Centre for Criminological Research.

Burnett, R. (2000) 'Understanding criminal careers through a series of in-depth interviews', *Offender Programs Report*, 4, 1, 1–2 and 13–16.

Burnett, R. and Maruna, S. (2004) 'So "prison works", does it? The criminal careers of 130 men released from prison under Home Secretary Michael Howard', *The Howard Journal*, 43, 4, 390–404.

Burt, C. (1925) *The Young Delinquent*, London: University of London Press.

Calverley, A., Cole, B., Kaur, G., Lewis, S., Raynor, P., Sadeghi, S., Smith, D., Vanstone, M. and Wardak, A. (2004) *Black and Asian Offenders on Probation*, Research study 277, London: Home Office.

Cann, J., Falshaw, L., Nugent, F. and Friendship, C. (2003) *Understanding What Works: Accredited Cognitive Skills Programmes for Adult Men and Young Offenders*, Home Office Research Findings 226, London: Home Office.

Carlen, P. (1989) 'Crime, inequality and sentencing', in P. Carlen and D. Cook (eds) *Paying for Crime*, Milton Keynes: Open University Press.

Carlen, P. (1994) 'Crime, inequality and sentencing', in A. Duff and D. Garland (eds) *A Reader on Punishment*, Oxford: Oxford University Press.

Carter, P. (2003) *Managing Offenders, Reducing Crime: A New Approach* (Correctional Services Review), London: Home Office.

Cavadino, M. and Dignan, J. (2002) *The Penal System: An Introduction* (third edition), London: Sage.

Cavadino, P. (1995) 'Gladstone: A century later', in Prison Reform Trust, *Gladstone at 100: Essays on the Past and Future of the Penal System*, London: Prison Reform Trust.

Chapman, T. and Hough, M. (1998) *Evidence-Based Practice*, London: Home Office.

Christie, N. (1993) *Crime Control as Industry*, London: Routledge.

Cicourel, A.V. (1968) *The Social Organisation of Juvenile Justice*, New York: Wiley.

Clear, T. and Karp, D. (1999) *The Community Justice Ideal: Preventing Crime and Achieving Justice*, Boulder, Colorado: Westview Press.

Clear, T.R., Rose, D.R. and Ryder, J.A. (2001) 'Incarceration and the community: The problem of removing and returning offenders', *Crime and Delinquency*, 47, 3, 335–67.

Coates, K. and Silburn, R. (1970) *Poverty: The Forgotten Englishmen*, Harmondsworth: Penguin.

Cohen, S. (1972) *Folk Devils and Moral Panics: The Creation of the Mods and Rockers*, London: MacGibbon and Kee.

Cohen, S. (1975) 'It's all right for you to talk: Political and sociological manifestoes for social work action', in R. Bailey and M. Brake (eds) *Radical Social Work*, London: Edward Arnold.

Cohen, S. (1979) 'The punitive city: Notes on the dispersal of social control', *Contemporary Crises*, 3, 339–63.

Cohen, S. (1985) *Visions of Social Control*, Cambridge: Polity Press.

Coleman, J.S. (1988) 'Social capital in the creation of human capital', *American Journal of Sociology* 94 (supplement), s95–s120.

Coleman, J.S. (1990) *Foundations of Social Theory*, London: Belknap Press.

Copas, J.B. (1992) *Statistical Analysis for A Risk of Reconviction Predictor*, Report to the Home Office, University of Warwick, unpublished.

Correctional Service of Canada (2003) *Circles of Support and Accountability: Guide to Project Development*, Ottawa: Correctional Service of Canada.

Correctional Services Accreditation Panel (2003) *Report 2002–2003*, London: CSAP.

Croft, J. (1978) *Research in Criminal Justice* (Home Office Research Study 44), London: HMSO.

Corrigan, P. and Leonard, P. (1978) *Social Work Practice Under Capitalism: A Marxist Approach*, London: Macmillan.

Cox, E.W. (1877) *The Principles of Punishment as Applied in the Administration of the Criminal Law by Judges and Magistrates*, London: Law Times Office.

Crow, I. (2001) *The Treatment and Rehabilitation of Offenders*, London: Sage.

Cullen, F.T. and Gilbert, K.E. (1982) *Reaffirming Rehabilitation*, Cincinnati, Ohio: Anderson Publishing.

Dalkin, J. and Padel, U. (2004) 'The reducing re-offending action plan and prisoner resettlement', *Criminal Justice Matters*, 56, 16–17.

Davies, A. Lloyd (1957) 'Psychotherapy and social casework II', in E.M. Goldberg, E. Irvine, A. Lloyd Davies and K. McDougall (eds) *The Boundaries of Casework*, London: Association of Psychiatric Social Workers.

Davies, M. (1969) *Probationers in their Social Environment*, Research Study, 2, London: HMSO.

Davies, M. (1974) *Social Work in the Environment*, Research Study, 21, London: HMSO.

Denman, G. (1982) *Intensive Intermediate Treatment with Juvenile Offenders: A Handbook of Assessment and Groupwork Practice*, Lancaster: Centre of Youth, Crime and Community, Lancaster University.

Departmental Committee on Prisons (1895) *Report from the Departmental Committee on Prisons* (The Gladstone Report), London: HMSO.

Dignan, J. (2005) *Understanding Victims and Restorative Justice*, Maidenhead: Open University Press.

Dignan, J. and Lowey, K. (2000) *Restorative Justice Options for Northern Ireland: A Comparative Review*, Belfast: Northern Ireland Office.

Dominelli, L. (1988) *Anti-Racist Social Work*, Basingstoke: Macmillan.

Dominelli, L. (2002) *Anti-Oppressive Social Work Theory and Practice*, Basingstoke: Palgrave Macmillan.

Dowden, C. and Andrews, D. (2004) 'The importance of staff practice in delivering effective correctional treatment: A meta-analysis', *International Journal of Offender Therapy and Comparative Criminology*, 48, 203–14.

Drakeford, M. (1983) 'Probation: Containment or liberty?', *Probation Journal*, 30, 7–10.

Drakeford, M. and Vanstone, M. (eds) (1996) *Beyond Offending Behaviour*, Aldershot: Arena.

Duff, R.A. (2001) *Punishment, Communication, and Community*, Oxford: Oxford University Press.

Duff, R.A. (2003) 'Probation, punishment and restorative justice: Should Al Truism be engaged in punishment?', *The Howard Journal*, 42, 2, 181–97.

Egan, G. (1986) *The Skilled Helper*, Monterey: Brooks-Cole.

Eglash, A. (1977) 'Beyond restitution – creative restitution', in J. Hudson and B. Galaway (eds) *Restitution in Criminal Justice: A Critical Assessment of Sanctions*, Lexington: Mass.: Heath and Company.

Ellis, T. (2000) 'Enforcement policy and practice: Evidence-based or rhetoric-based?', *Criminal Justice Matters*, 39, 6–8.

Eshelby, S. (1962) 'The Probation Officer as caseworker', *British Journal of Psychiatric Social Work*, 6, 125–30.

Falshaw, L., Friendship, C., Travers, R. and Nugent, F. (2003) *Searching for 'What Works': An Evaluation of Cognitive Skills Programmes*, Home Office Research Findings 206, London: Home Office.

Farrall, S. (2002) *Rethinking What Works with Offenders: Probation, Social Context and Desistance from Crime*, Cullompton: Willan.

Farrall, S. and Bowling, B. (1999) 'Structuration, human development and desistance from crime', *British Journal of Criminology*, 39, 2, 253–68.

Farrall, S. and Maruna, S. (2004) 'Desistance-focused criminal justice policy research', *The Howard Journal*, 43, 4, 358–67.

Faulkner, D. (2003) 'Taking citizenship seriously: Social capital and criminal justice in a changing world', *Criminal Justice*, 3, 3, 287–315.

Faulkner, D. (2004) 'A moment of opportunity: Some reflections on the Carter Report and on the Government's accompanying statement "Reducing Crime: Changing Lives"', *VISTA*, 9, 1, 2–8.

Faulkner, D. and Flaxington, F. (2004) 'NOMS and Civil Renewal', *VISTA*, 9, 2, 90–99.

Feeley, M. and Simon, J. (1992) 'The new penology: Notes on the emerging strategy of corrections and its implications', *Criminology*, 30, 449–74.

Feeley, M. and Simon, J. (1994) 'Actuarial justice: The emerging new criminal law', in D. Nelken (ed.) *The Futures of Criminology*, London: Sage.

Feilzer, M., Appleton, C., Roberts, C. and Hoyle, C. (2004) *Cognitive Behaviour Projects: The National Evaluation of the Youth Justice Board's Cognitive Behaviour Projects*, London: Youth Justice Board.

Fielding, H. (1751) 'An enquiry into the causes of the late increase of robbers', reproduced in M.R. Zirker (ed.) (1988) *Henry Fielding: An Enquiry into the Causes of the Late Increase of Robbers and Related Writings*, Oxford: Clarendon Press.

Fielding, H. (1753) 'A Proposal for making effectual provision for the poor', reproduced in M.R. Zirker (ed.) (1988) *Henry Fielding: An Enquiry into the Causes of the Late Increase of Robbers and Related Writings*, Oxford: Clarendon Press.

Fischer, J. (1973) 'Is casework effective? A review', *Social Work*, 18, 5–20.

Fischer, J. (1976) *The Effectiveness of Social Casework*, Springfield: C.C. Thomas.

Fischer, J. (1978) *Effective Casework Practice*, New York: McGraw-Hill.

Fischer, J. and Gochros, H. (1975) *Planned Behaviour Change: Behaviour Modification in Social Work*, New York: Free Press.

Flew, A. (1973) *Crime or Disease?* London: Macmillan.

Folkard, M.S., Fowles, A.J., McWilliams, B.C., McWilliams, W., Smith, D.D. and Walmsley, G.R. (1974) *IMPACT. Intensive Matched Probation and After-care Treatment: Volume I. The Design of the Experiment and An Interim Evaluation*, Home Office Research Study 24, London: HMSO.

Folkard, M.S., Smith, D.E. and Smith, D.D. (1976) *IMPACT Volume II: The Results of the Experiment*, Home Office Research Study 36, London: HMSO.

Foren, R. and Bailey, R. (1968) *Authority in Social Casework*, Oxford: Pergamon Press.

Forsythe, W.J. (1991) *Penal Discipline, Reformatory Projects and the English Prison Commission 1895–1939*, Exeter: University of Exeter Press.

Foucault, M. (1977) *Discipline & Punish*, London: Allen Lane.

Fowles, A.J. (1978) *Prison Welfare: An Account of an Experiment at Liverpool*, Home Office Research Study 45, London: HMSO.

Friendship, C., Blud, L., Erikson, M. and Travers, R. (2002) *An Evaluation of Cognitive Behavioural Treatment for Prisoners*, Home Office Research Findings 161, London: Home Office.

Friendship, C., Mann, R. and Beech, A. (2003) *The Prison-based Sex Offender Treatment Programme – an evaluation*, Research Findings 205, London: Home Office.

Frude, N., Honess, T. and Maguire, M. (1994) *CRIME-PICS II Manual*, Cardiff: Michael and Associates.

Gaes, G., Flanagan, T., Motiuk, L. and Stewart, L. (1999) 'Adult correctional treatment', in M. Tonry and J. Petersilia (eds) *Crime and Justice: A Review of Research*, 26, 361–426, Chicago: University of Chicago Press.

Garland, D. (1985) *Punishment and Welfare: A History of Penal Strategies*, Aldershot: Gower.

Garland, D. (1995) 'Penal modernism and postmodernism', in T.G. Blomberg and S. Cohen (eds) *Punishment and Social Control*, New York: Aldine de Gruyter.

Garland, D. (1996) 'The limits of the sovereign state: Strategies of crime contro in contemporary society', *British Journal of Criminology*, 36, 4, 445–71.

Garland, D. (1997a) 'Probation and the reconfiguration of crime control', in R. Burnett (ed.) *The Probation Service: Responding to Change* (Proceedings of the Probation Studies Unit First Colloquium: Probation Studies Unit Report No. 3), Oxford: University of Oxford Centre for Criminological Research.

Garland, D. (1997b) ' "Governmentality" and the problem of crime: Foucault, criminology, sociology', *Theoretical Criminology*, 1, 2, 173–214.

Garland, D. (2000) 'The culture of high crime societies: Some preconditions of recent "Law and Order" policies', *British Journal of Criminology*, 40, 3, 347–75.

Garland, D. (2001) *The Culture of Control*, Oxford: Oxford University Press.

Gendreau, P. and Ross, R. (1980) 'Effective correctional treatment: Bibliotherapy for cynics', in R. Ross and P. Gendreau (eds) *Effective Correctional Treatment*, Toronto: Butterworths.

Gendreau, P., Goggin, C. and Smith, P. (1999) 'The forgotten issue in effective correctional treatment: Program implementation', *International Journal of Offender Therapy and Comparative Criminology*, 43, 180–87.

Ghate, D. and Ramella, M. (2002) *Positive Parenting: The National Evaluation of the Youth Justice Board's Parenting Programme*, London: Youth Justice Board.

Gibran, K. (1991) *The Prophet*, London: Pan.

Giddens, A. (1990) *The Consequences of Modernity*, Cambridge: Polity Press.

Giddens, A. (1991) *Modernity and Self-Identity*, Cambridge: Polity Press.

Glueck, S. and Glueck, E. (1934) *One Thousand Juvenile Delinquents*, Cambridge, Mass.: Harvard University Press.

Glueck, S. and Glueck, E. (1940) *Juvenile Delinquents Grown Up*, New York: The Commonwealth Fund.

Goffman, E. (1961) *Asylums*, Garden City: Anchor Books.

Goldberg, E.M., Gibbons, J. and Sinclair, I. (1985) *Problems, Tasks and Outcomes*, London: Allen and Unwin.

Goldberg, E.M., Irvine, E., Lloyd Davies, A. and McDougall, K. (eds) (1957) *The Boundaries of Casework*, London: Association of Psychiatric Social Workers.

Gostin, L. (1977) *A Human Condition*, London: MIND.

Gottfredson, M.R. and Hirschi, T. (1990) *A General Theory of Crime*, California: Stanford University Press.

Gottfredson, M.R. and Hirschi, T. (1995) 'National crime control policies', *Society*, 32, 30–36.

Grey, A. and Dermody, H. (1972) 'Reports of casework failure', *Social Casework*, November, 534–43.

Grunhut, M. (1952) 'Probation in Germany', *Howard Journal*, 8, 168–74.

Hall, M.P. (1952) *The Social Services of Modern England*, London: Routledge.

Hall, P. (1976) *Reforming the Welfare*, London: Heinemann.

Halliday, J. (2001) *Making Punishments Work: Report of a Review of the Sentencing Framework for England and Wales*, London: Home Office.

Hammersley, R., Reid, M., Oliver, A., Genova, A., Raynor, P., Minkes, J. and Morgan, M. (2004) *Drug and Alcohol Projects: The National Evaluation of the Youth Justice Board's Drug and Alcohol Projects*, London: Youth Justice Board.

Harper, G. and Chitty, C. (2004) *The Impact of Corrections on Re-offending: A Review of 'What Works'*, Home Office Research Study 291, London: Home Office.

Harper, R. and Hardy, S. (2000) 'An evaluation of motivational interviewing as a method of intervention with clients in a probation setting', *British Journal of Social Work*, 30, 393–400.

Haslewood-Pocsik, I., Merone, L. and Roberts, C. (2004) *The Evaluation of the Employment Pathfinder: Lessons from Phase 1 and A Survey for Phase 2*. Online Report 22/04, London: Home Office.

Haxby, D. (1978) *Probation*, London: Constable.

Heath, B., Raynor, P. and Miles, H. (2002) 'What Works in Jersey: The first ten years', *VISTA*, 7, 202–208.

Hearnden, I. and Millie, A. (2003) *Investigating Links between Probation Enforcement and Reconviction*, Online Report 41/03, London: Home Office.

Hedderman, C. (2004) 'Testing times: How the policy and practice environment shaped the creation of the What Works evidence-base', *VISTA*, 8, 182–88.

Hedderman, C. and Hearnden, I. (2001) 'To discipline or punish? Enforcement under National Standards 2000', *VISTA*, 6, 215–24.

Hine, J. and Celnick, A. (2001) *A One-year Reconviction Study of Final Warnings*, Sheffield: University of Sheffield.

Hirschi, T. (1969) *Causes of Delinquency*, Berkeley: California University Press.

HM Inspectorate of Probation (1995) *Dealing with Dangerous People: The Probation Service and Public Protection*, London: Home Office.

HM Inspectorate of Probation (1997) 'Risk management guidance', in Home Office *Management and Assessment of Risk in the Probation Service*, London: Home Office.

HM Inspectorate of Probation (1998) *Exercising Constant Vigilance: The Role of the Probation Service in Protecting the Public from Sex Offenders*, London: Home Office.

HM Inspectorate of Probation (2001) *Through the Prison Gate: A Joint Thematic Review by HM Inspectorates of Prison and Probation*, London: Home Office.

Holdaway, S., Davidson, N., Dignan, J., Hammersley, R., Hine, J. and Marsh, P. (2001) *New Strategies to Address Youth Offending: The National Evaluation of the Pilot Youth Offending Teams*, RDS Occasional Paper 69, London: Home Office.

Hollin, C. (2001) 'Rehabilitation', in E. McLaughlin and J. Muncie (eds) *The Sage Dictionary of Criminology*, London: Sage.

Hollin, C., McGuire, J., Palmer, E., Bilby, C., Hatcher, R. and Holmes, A. (2002) *Introducing Pathfinder Programmes into the Probation Service: An Interim Report*, Home Office Research Study 247, London: Home Office.

Hollin, C., Palmer, E., McGuire, J., Hounsome, J., Hatcher, R., Bilby, C. and Clark, C. (2004) *Pathfinder Programmes in the Probation Service: A Retrospective Analysis*, Home Office Online Report 66/04, London: Home Office.

Hollis, F. (1964) *Casework: A Psychosocial Therapy*, New York: Random House.

Home Office (1963) *The Organisation of After-Care: Report of the Advisory Council on the Treatment of Offenders*, London: HMSO.

Home Office (1966) *Report of the Inquiry into Prison Escapes and Security* (Mountbatten Report), Cmnd 3175, London: HMSO.

Home Office (1969) *People in Prison*, Cmnd 4214, London: HMSO.

Home Office (1977) *A Review of Criminal Justice Policy 1976*, London: HMSO.

Home Office (1984) *Probation Service in England and Wales: Statement of National Objectives and Priorities*, London: Home Office.

Home Office (1988) *Tackling Offending: An Action Plan*, London: Home Office.

Home Office (1990) *Crime, Justice and Protecting the Public*, Cmnd 965, London: HMSO.

Home Office (1991) *Partnership in Dealing with Offenders in the Community: A Decision Document*, London: Home Office.

Home Office (1992) *National Standards for the Supervision of Offenders in the Community*, London: Home Office.

Home Office (1995a) *Incident Reporting* (Probation Circular 41/1995), London: Home Office.

Home Office (1995b) *National Standards for the Supervision of Offenders in the Community*, London: Home Office.

Home Office (1995c) *Managing What Works: Conference Report and Guidance on Critical Success Factors for Probation Supervision Programmes*, Probation Circular 77/1995, London: Home Office.

Home Office (1996a) *Protecting the Public*, Cmnd 3190, London: HMSO.

Home Office (1996b) *Guidance for the Probation Service on the Offender Group Reconviction Scale (OGRS)*, (Probation Circular 63/1996), London: Home Office.

Home Office (1998) *Joining Forces to Protect the Public: Prisons-Probation*, London: Home Office.

Home Office (2001) *Through the Prison Gate: A Joint Thematic Review by HM Inspectorates of Prisons and Probation*, London: Home Office.

Home Office (2003) *Restorative Justice: The Government's Strategy*, London: Home Office.

Home Office (2004a) *Reducing Crime, Changing Lives*, London: Home Office.

Home Office (2004b) *Reducing Re-offending National Action Plan*, London: Home Office.

Homel, P., Nutley, S., Webb, B. and Tilley, N. (2005) *Investing to Deliver: Reviewing the Implementation of the UK Crime Reduction Programme*, Home Office Research Study 281, London: Home Office.

Hood, R. (1974) *Tolerance and the Tariff*, London: NACRO.

Hood, R. (2004) 'Hermann Mannheim and Max Grunhut: Criminological pioneers in London and Oxford', *British Journal of Criminology*, 44, 4, 469–95.

Hough, M. (ed.) (2004) *Criminal Justice* 4, 3. Special Issue: *Evaluating the Crime Reduction Programme in England and Wales*.

Howard, J. (1929) [1777] *The State of the Prisons*, London: J.M. Dent and Sons.

Howard, M. (1993) Speech to Conservative Party conference, October.

Hudson, B. (1987) *Justice Through Punishment: A Critique of the 'Justice' Model of Corrections*, Basingstoke: Macmillan Education.

Hudson, B. (1988) 'Social skills training in practice', *Probation Journal*, 35, 85–91.

Hudson, B. (1999) 'Punishment, poverty and responsibility: The case for a hardship defence', *Social and Legal Studies*, 8, 4, 583–91.

Hudson, B. (2003) *Understanding Justice*, second edition, Buckingham: Open University Press.

Hunt, A.W. (1964) 'Enforcement in probation casework', *British Journal of Delinquency*, 4, 239–52.

Ignatieff, M. (1978) *A Just Measure of Pain: The Penitentiary in the Industrial Revolution*, New York: Pantheon Books.

James, A. (1995) 'Probation values for the 1990s – and beyond?', *Howard Journal*, 34, 4, 326–43.

Jarvis, F. (1972) *Advise, Assist and Befriend: A History of the Probation and After-Care Service*, London: National Association of Probation Officers.

Johnstone, G. (2002) *Restorative Justice: Ideas, Values, Debates*, Cullompton: Willan.

Jones, R. (1984) 'Questioning the new orthodoxy', *Community Care*, 11 October, 26–29.

Jordan, B. (2000) *Social Work and the Third Way*, London: Sage.

JUSTICE, The Howard League for Penal Reform and the National Association for the Care and Resettlement of Offenders (1972), *Living It Down: The Problem of Old Convictions*, London: Stevens.

Kant, I. (1965) [1788] *Critique of Practical Reason* (trans. L. Beck), Indianapolis: Bobbs-Merrill.

Kemshall, H. (1998) *Risk in Probation Practice*, Aldershot: Ashgate.

Kemshall, H. and Canton, R. (2002) *The Effective Management of Programme Attrition*, London: National Probation Service.

Kemshall, H. and Maguire, M. (2003) 'Sex offenders, risk penality and the problem of disclosure to the community', in A. Matravers (ed.) *Sex Offenders in the Community*, Cullompton: Willan.

Kemshall, H., Canton, R. and Bailey, R. (2004) 'Dimensions of difference', in A. Bottoms, S. Rex and G. Robinson (eds) *Alternatives to Prison: Options for an Insecure Society*, Cullompton: Willan.

Kemshall, H., Parton, N., Walsh, M. and Waterson, J. (1997) 'Concepts of risk in relation to organizational structure and functioning within the personal social services and probation', *Social Policy and Administration*, 31, 213–32.

Kendall, K. (2004) 'Dangerous thinking: A critical history of correctional cognitive behaviouralism', in G. Mair (ed.) *What Matters in Probation*, Cullompton: Willan.

Kent Probation and After-Care Service (1981) 'Probation Control Unit: A community-based experiment in intensive supervision', in *Annual Report on the Work of the Medway Centre*, Maidstone: Kent Probation and After-Care Service.

King, R.D. and Elliott, K.W. (1978) *Albany: Birth of a Prison – End of an Era*, London: Routledge and Kegan Paul.

King, R.D. and Morgan, R. (1980) *The Future of the Prison System*, Farnborough: Gower.

Laub, J.H. and Sampson, R.J. (2001) 'Understanding desistance from crime', in M. Tonry (ed.) *Crime and Justice: A Review of Research*, 28, Chicago: University of Chicago Press.

Lemert, E. (1967) *Human Deviance, Social Problems and Social Control*, New Jersey: Prentice-Hall.

Leonard, P. (1975) 'Towards a paradigm for radical practice', in R. Bailey and M. Brake (eds) *Radical Social Work*, London: Edward Arnold.

Lewis, C. (2004) 'Trends in crime, victimisation and punishment', in A. Bottoms, S. Rex and G. Robinson (eds) *Alternatives to Prison: Options for an Insecure Society*, Cullompton: Willan.

Lewis, C.S. (1971) 'The humanitarian theory of punishment', in *Undeceptions*, London: Geoffrey Bles.

Lewis, S., Vennard, J., Maguire, M., Raynor, P., Vanstone, M., Raybould, S. and Rix, A. (2003) *The Resettlement of Short-term Prisoners: An Evaluation of Seven Pathfinders*, RDS Occasional Paper No. 83, London: Home Office.

Lipsey, M. (1992) 'Juvenile delinquency treatment: A meta-analytic enquiry into the variability of effects', in T. Cook, H. Cooper, D.S. Cordray, H. Hartmann, L.V. Hedges, R.L. Light, T.A. Louis and F. Mosteller (eds) *Meta-Analysis for Explanation: A Case-Book*, New York: Russell Sage, pp. 83–127.

Lipsey, M. (1999) 'Can rehabilitative programs reduce the recidivism of juvenile offenders? An inquiry into the effectiveness of practical programs', *Virginia Journal of Social Policy and the Law*, 6, 611–41.

Lipsey, M. and Wilson, D. (1998) 'Effective intervention for serious juvenile offenders', in R. Loeber and D. Farrington (eds) *Serious and Violent Juvenile Offenders: Risk Factors and Successful Interventions*, Thousand Oaks: Sage.

Lipton, D., Martinson, R. and Wilks, J. (1975) *The Effectiveness of Correctional Treatment*, New York: Praeger.

Lipton, D., Pearson, F., Cleland, C. and Yee, D. (2002) 'The effectiveness of cognitive-behavioural treatment methods on offender recidivism', in J. McGuire (ed.) *Offender Rehabilitation and Treatment*, Chichester: Wiley.

Lobley, D., Smith, D. and Stern, C. (2001) *Freagarrach: An Evaluation of a Project for Persistent Juvenile Offenders*, Edinburgh: Scottish Executive Central Research Unit.

Loney, M. (1983) *Community against Government: The British Community Development Project 1968–78*, London: Heinemann.

Lopez-Rey, M. (1957) 'United Nations activities and international trends in probation', *Howard Journal*, 9, 346–53.

Lovbakke, J. and Homes, A. (2004) *Focus on Female Offenders: The Real Women Programme – Probation Service Pilot*, Development and Practice Report 18, London: Home Office.

Macadam, E. (1925) *The Equipment of the Social Worker*, London: Allen and Unwin.

Macadam, E. (1945) *The Social Servant in the Making*, London: Allen and Unwin.

Macpherson, G. (1992) *Black's Medical Dictionary* (thirty-seventh edition), London: A and C Black Publishers.

Maguire, M. (2004a) 'The crime reduction programme in England and Wales: Reflections on the vision and the reality', *Criminal Justice*, 4, 3, 213–37.

Maguire, M. (2004b) 'Resettlement of short-term prisoners: Some new approaches', *Criminal Justice Matters*, 56, 22–23.

Maguire, M. and Raynor, P. (1997) 'The revival of throughcare: Rhetoric and reality in Automatic Conditional Release', *British Journal of Criminology*, 37, 1, 1–14.

Maguire, M., Raynor, P., Vanstone, M. and Kynch, J. (2000) 'Voluntary after-care and the probation service: A case of diminishing responsibility', *Howard Journal of Criminal Justice*, 39, 234–48.

Mair, G. (1994) 'Standing at the crossroads: What works in community penalties?', paper to the National Conference for Probation Committee Members, Scarborough.

Mair, G. (1997) 'Community penalties and the probation service', in M. Maguire, R. Morgan and R. Reiner (eds) *The Oxford Handbook of Criminology* (second edition), Oxford: Clarendon Press.

Mair, G. (ed.) (2004) *What Matters in Probation*, Cullompton: Willan.

Mair, G., Lloyd, C., Nee, C. and Sibbitt, R. (1994) *Intensive Probation in England and Wales: An Evaluation*, Home Office Research Study 133, London: HMSO.

Mannheim, H. (1939) *The Dilemma of Penal Reform*, London: Allen and Unwin.

Mannheim, H. (1946) *Criminal Justice and Social Reconstruction*, London: Routledge.

Marshall, T.F. (1999) *Restorative Justice: An Overview*, London: Home Office.

Martinson, R. (1974) 'What works? Questions and answers about prison reform', *The Public Interest*, 35, 22–54.

Martinson, R. (1979) 'New findings, new views: A note of caution regarding sentencing reform', *Hofstra Law Review*, 7, 243–58.

Maruna, S. (2000) *Making Good*, Washington: American Psychological Association.

Maruna, S. (2004) ' "California Dreamin": Are we heading toward a national offender "waste management" service?', *Criminal Justice Matters*, 56, 6–7.

Maruna, S. and LeBel, T. (2002) 'Revisiting ex-prisoner re-entry: A buzz-word in search of a narrative', in S. Rex and M. Tonry (eds) *Reform and Punishment: The Future of Sentencing*, Cullompton: Willan.

Maruna, S. and LeBel, T. (2003) 'Welcome home? Examining the "re-entry court" concept from a strengths-based perspective', *Western Criminology Review*, 4, 91–107. http://wcr.sonoma.edu.

Maruna, S., Immarigeon, R. and LeBel, T.P. (2004) 'Ex-offender reintegration: Theory and practice', in S. Maruna and R. Immarigeon (eds) *After Crime and Punishment: Pathways to Offender Reintegration*, Cullompton: Willan.

Mathiesen, T. (1983) 'The future of control systems – the case of Norway', in D. Garland and P. Young (eds) *The Power To Punish: Contemporary Penality and Social Analysis*, London: Heinemann Educational Books.

Maxwell, G. and Morris, A. (1993) *Family, Victims and Culture: Youth Justice in New Zealand*, Wellington: Social Policy and Administration and Victoria University of Wellington.

Maxwell, G. and Morris, A. (2004) 'What is the place of shame in restorative justice?', in H. Zehr and B. Toews (eds) *Critical Issues in Restorative Justice*, Monsey, NY: Criminal Justice Press.

May, C. and Wadwell, J. (2001) *Enforcing Community Penalties: The Relationship between Enforcement and Reconviction*, Research Findings 155, London: Home Office.

May, Mr Justice (Chairman) (1979) *Report of the Committee of Inquiry into the United Kingdom Prison Services*, Cmnd 7673, London: HMSO.

Mayer, J. and Timms, N. (1970) *The Client Speaks*, London: Routledge.

Mays, J.B. (1957) 'Social research and social casework II', in E.M Goldberg, E. Irvine, A. Lloyd Davies and K. McDougall (eds) *The Boundaries of Casework*, London: Association of Psychiatric Social Workers.

McConville, S. (1981) *A History of English Prison Administration, Volume 1: 1750–1877*, London: Routledge and Kegan Paul.

McConville, S. (1995a) 'The Victorian prison', in N. Morris and D.J. Rothman (eds) *The Oxford History of the Prison*, Oxford: Oxford University Press.

McConville, S. (1995b) *English Local Prisons 1860–1900: Next only to Death*, London: Routledge.

McCord, J. (1978) 'A thirty-year follow-up of treatment effects', *American Psychologist*, 33, 284–89.

McGuire, J. (ed.) (1995) *What Works: Reducing Reoffending*, Chichester: Wiley.

McGuire, J. (2000) *Cognitive-Behavioural Approaches*, London: Home Office.

McGuire, J. (2002) 'Integrating findings from research reviews', in J. McGuire (ed.) *Offender Rehabilitation and Treatment*, Chichester: Wiley.

McGuire, J. (ed.) (2002) *Offender Rehabilitation and Treatment: Effective Programmes and Policies to Reduce Re-offending*, Chichester: Wiley.

McGuire, J. (2003) 'Maintaining change: Converging legal and psychological initiatives in a therapeutic jurisprudence framework', *Western Criminology Review*, 4, 108–23. http://wcr.sonoma.edu.

McGuire, J. and Priestley, P. (1985) *Offending Behaviour: Skills and Stratagems for Going Straight*, London: Batsford.

McIvor, G. (1990) *Sanctions for Serious or Persistent Offenders*, Stirling: Social Work Research Centre.

McIvor, G., Eley, S., Malloch, M. and Yates, R. (2003) *Establishing Drug Courts in Scotland: Early Experiences of the Pilot Drug Courts in Glasgow and Fife*, Research Findings 71/2003, Edinburgh: Scottish Executive.

McIvor, G., Jamieson, J., Gayle, V., Moodie, K. and Netten, A. (2000) *Evaluation of the Airborne Initiative (Scotland)*, Edinburgh: Scottish Executive Central Research Unit.

McLaren, K. (1992) *Reducing Reoffending: What Works Now?*, Wellington, N.Z.: Department of Justice.

McMahon, G., Hall, A., Hayward, G., Hudson, C. and Roberts, C. (2004) *Basic Skills Programmes in the Probation Service: An Evaluation of the Basic Skills Pathfinder*, Home Office Research Findings 203, London: Home Office.

McNeill, F. (2003) 'Desistance focused probation practice', in W.-H. Chui and M. Nellis (eds) *Moving Probation Forward*, Harlow: Pearson Education.

McNeill, F. (2004) 'Desistance, rehabilitation and correctionalism: Developments and prospects in Scotland', *The Howard Journal*, 43, 4, 420–36.

McWilliams, W. (1983) 'The mission to the English Police Courts 1876–1936', *The Howard Journal*, 22, 129–47.

McWilliams, W. (1987) 'Probation, pragmatism and policy', *The Howard Journal*, 26, 97–121.

McWilliams, W. (1989) 'An expressive model for evaluating probation practice', *Probation Journal*, 36, 58–64.

McWilliams, W. and Pease, K. (1990) 'Probation practice and an end to punishment', *The Howard Journal*, 29, 14–24.

Mead, G.H. (1934) *Mind, Self and Society*, Chicago: University of Chicago Press.

Merrington, S. and Stanley, S. (2000) 'Doubts about the What Works Initiative', *Probation Journal*, 47, 272–75.

Merrington, S. and Stanley, S. (2004) 'What Works? Revisiting the evidence in England and Wales', *Probation Journal*, 51, 7–20.

Meyer, H., Borgatta, E. and Jones, W. (1965) *Girls at Vocational High*, New York: Russell Sage Foundation.

Miers, D. (2004) 'Situating and researching restorative justice in Great Britain', *Punishment and Society*, 6, 1, 23–46.

Miers, D., Maguire, M., Goldie, S., Sharpe, K., Hale, C., Netton, A., Uglow, S., Doolin, K., Hallam, A., Enterkin, J. and Newburn, T. (2001) *An Exploratory Evaluation of Restorative Schemes*, Crime Reduction Research Series Paper 9. London: Home Office.

Miles, A.P. (1954) *American Social Work Theory*, New York: Harper.

Miles, H. (2004) 'The Parish Hall Enquiry: A community-based alternative to formal court processing in the Channel Island of Jersey', *Probation Journal*, 51, 2, 133–44.

Miles, H. and Raynor, P. (2004) *Community Sentences in Jersey: Risks, Needs and Rehabilitation*, St Helier: Jersey Probation and After-Care Service.

Miller, W.R. and Rollnick, S. (eds) (1992) *Motivational Interviewing: Preparing People to Change Addictive Behavior*, New York: Guilford Press.

Mills, C.W. (1943) 'The professional ideology of social pathologists', *American Journal of Sociology*, 49, 165–80.

Morgan Harris Burrows (2003) *Evaluation of the Youth Inclusion Programme: End of Phase One Report*, London: Youth Justice Board.

Morgan, R. (2002) 'Something has got to give', *HLM – The Howard League Magazine*, 20, 4, 7–8.

Morgan, R. (2003) 'Foreword', *Her Majesty's Inspectorate of Probation Annual Report 2002/2003*, London: Home Office.

Morris, A., Giller, H., Szwed, E. and Geach, H. (1980) *Justice for Children*, London: Macmillan.

Mowat, C.L. (1961) *The Charity Organisation Society 1869–1913*, London: Methuen.

Mullen, E. and Dumpson, J. (1972) 'Is social work on the wrong track?', in Mullen, Dumpson and associates, *Evaluation of Social Intervention*, London: Jossey-Bass.

Muncie, J. (2001) 'Prison histories: Reform, repression and rehabilitation', in E. McLaughlin and J. Muncie (eds) *Controlling Crime* (second edition), London: Sage.

Murray, C. (1990) *The Emerging British Underclass*, London: IEA Health and Welfare Unit.

NACRO (2000) *The Forgotten Majority: The Resettlement of Short-term Prisoners*, London: NACRO.

NAPO (2001) 'AGM resolutions 2001', *NAPO News*, 134, 10–15.

NAPO (2005) 'NOMS – heading for the rocks?', *NAPO News*, 167, 1–2.

National Probation Service (2001) *A New Choreography*, London: Home Office.

National Probation Service (2004) *Accredited Programmes Performance Report 2002–3*, London: National Probation Service.

Nellis, M. (1990) 'Probation, the state and the independent sector', in P. Senior and D. Woodhill (eds) *Criminal Justice in the 1990s: What Future(s) for the Probation Service?*, Sheffield: PAVIC Publications from Sheffield City Polytechnic.

Nellis, M. (1995) 'Probation values for the 1990s', *The Howard Journal*, 34, 19–44.

Nellis, M. (2004) review of *Understanding Community Penalties*, by P. Raynor and M. Vanstone, 2002, in *Punishment and Society*, 5, 4, 478–81.

Nellis, M. (2004) 'Into the field of corrections: The end of Englash probation in the early 21st century?', *Cambrian Law Review*, 35, 115–33.

Newburn, T., Crawford, A., Earle, R., Goldie, S., Hale, C., Netten, A., Saunders, R., Hallam, A., Sharpe, K. and Uglow, S. (2002) *The Introduction of Referral Orders into the Youth Justice System*, Research Study 242, London: Home Office.

OASys Development Team (2001) *Offender Assessment System User Manual*, London: Home Office.

O'Malley, P. (1992) 'Risk, power and crime prevention', *Economy and Society*, 21, 252–75.

O'Malley, P. (2001) 'Crime, risk and prudentialism revisited', in K. Stenson and R.R. Sullivan (eds) *Crime, Risk and Justice*, Devon: Willan Publishing.

Ong, G., Harsent, L. and Coles, S. (2003) *Think First Evaluation*, Unpublished conference workshop report, London: National Probation Service.

Palmer, T. (1974) 'The Youth Authority's Community Treatment Project', *Federal Probation*, 38, 3–14.

Palmer, T. (1975) 'Martinson revisited', *Journal of Research in Crime and Delinquency*, 12, 133–52.

Pansegrouw, N.J. de W. (1952) 'Probation and its place in a rational and humane programme for the treatment of offenders', in Report of the European Seminar on Probation, London 20–30 October 1952: United Nations (unpublished).

Partridge, S. (2004) *Examining Case Management Models for Community Sentences*, Home Office Online Report 17/04, London: Home Office.

Pease, K. (1980) 'The future of the community treatment of offenders in Britain', in A.E. Bottoms and R.H. Preston (eds) *The Coming Penal Crisis*, Edinburgh: Scottish Academic Press.

Pease, K. (1999) 'The probation career of Al Truism', *The Howard Journal*, 38, 1, 2–16.

Pease, K., Billingham, S. and Earnshaw, I. (1977) *Community Service Assessed in 1976*, Home Office Research Study No. 39, London: HMSO.

Petersilia, J. (1990) 'Conditions that permit intensive supervision programmes to survive', *Crime and Delinquency*, 36, 126–45.

Petersilia, J. (2003) *When Prisoners Come Home: Parole and Prisoner Reentry*, Oxford: Oxford University Press.

Petrunik, M.G. (2002) 'Managing unacceptable risk: Sex offenders, community response, and social policy in the United States and Canada', *International Journal of Offender Therapy and Comparative Criminology*, 46, 4, 483–511.

Pettes, D. (1967) *Supervision in Social Work*, London: Allen and Unwin.

Pincus, A. and Minahan, A. (1973) *Social Work Practice: Model and Method*, Itasca: Peacock.

Pitts, J. (1988) *The Politics of Juvenile Crime*, London: Sage.

Plant, R. (1970) *Social and Moral Theory in Casework*, London: Routledge.

Ploeg, G.J. (2003) 'Moving probation forward: A review', *Bulletin of the Conference Permanente Europeenne de la Probation*, 29, 8.

Porporino, F., Van Dieten, M. and Fabiano, E. (2003) *A Women's Programme for Acquisitive Female Offenders: Theory and Applications Manual*, Ottawa: T3 Associates.

Powers, E. and Witmer, H. (1951) *An Experiment in the Prevention of Delinquency*, New York: Columbia University Press.

Priestley, P., McGuire, J., Flegg, D., Hemsley, V. and Welham, D. (1978) *Social Skills and Personal Problem Solving: A Handbook of Methods*, London: Tavistock.

Priestley, P., McGuire, J., Flegg, D., Hemsley, D., Welham, D. and Barnitt, R. (1984) *Social Skills in Prisons and in the Community*, London: Routledge and Kegan Paul.

Pugh, R.B. (1970) *Imprisonment in Medieval England*, Cambridge: Cambridge University Press.

Radzinowicz, L. (ed.) (1958) *The Results of Probation*, A Report of the Cambridge Department of Criminal Science, London: Macmillan.

Radzinowicz, L. and Hood, R. (1990) *The Emergence of Penal Policy*, Oxford: Clarendon Press.

Rawls, J. (1972) *A Theory of Justice*, Oxford: Oxford University Press.

Raynor, P. (1978) 'Compulsory persuasion: A problem for correctional social work', *British Journal of Social Work*, 8, 4, 411–24.

Raynor, P. (1985) *Social Work, Justice and Control*, Oxford: Blackwell.

Raynor, P. (1988) *Probation as an Alternative to Custody*, Aldershot: Avebury.

Raynor, P. (1995) ' "What works": Probation and forms of justice', in D. Ward and M. Lacey (eds) *Probation: Working for Justice*, London: Whiting and Birch.

Raynor, P. (1996) 'Evaluating Probation: The rehabilitation of effectiveness', in T. May and A. Vass (eds) *Working with Offenders*, London: Sage.

Raynor, P. (1997) 'Evaluating probation: A moving target', in G. Mair (ed.) *Evaluating the Effectiveness of Community Penalties*, Aldershot: Avebury.

Raynor, P. (1998a) 'Reading probation statistics: A critical comment', *VISTA*, 3, 181–85.

Raynor, P. (1998b) 'Attitudes, social problems and reconvictions in the STOP probation experiment', *Howard Journal*, 37, 1–15.

Raynor, P. (2001) 'Community penalties and social integration: "Community" as solution and as problem', in A. Bottoms, L. Gelsthorpe and S. Rex (eds) *Community Penalties: Change and Challenges*, Cullompton: Willan.

Raynor, P. (2003) 'Evidence-based probation and its critics', *Probation Journal*, 50, 334–45.

Raynor, P. (2004a) 'The probation service 'pathfinders': Finding the path and losing the way?', *Criminal Justice*, 4, 3, 309–25.

Raynor, P. (2004b) 'Opportunity, motivation and change: Some findings from research on resettlement', in R. Burnett and C. Roberts (eds) *What Works in Probation and Youth Justice*, Cullompton: Willan.

Raynor, P. and Vanstone, M. (1984) 'Putting practice into theory – an in-college skills training programme', *Issues in Social Work Education*, 4, 85–93.

Raynor, P. and Vanstone, M. (1996) 'Reasoning and Rehabilitation in Britain: The results of the Straight Thinking On Probation (STOP) programme', *International Journal of Offender Therapy and Comparative Criminology*, 40, 272–84.

Raynor, P. and Vanstone, M. (1997) *Straight Thinking On Probation (STOP): The Mid Glamorgan Experiment. Probation Studies Unit Report No. 4*, Oxford: University of Oxford Centre for Criminological Research.

Raynor, P. and Vanstone, M. (2001) *Understanding Community Penalties*, Buckingham: Open University Press.

Raynor, P. and Vanstone, M. (2002) 'Straight thinking on probation: Evidence-based practice and the culture of curiosity', in G. Bernfeld, D. Farrington and A. Leschied (eds) *Offender Rehabilitation in Practice*, Chichester: Wiley.

Raynor, P., Smith, D. and Vanstone, M. (1994) *Effective Probation Practice*, Basingstoke: Macmillan.

Raynor, P., Kynch, J., Roberts, C. and Merrington, M. (2000) *Risk and Need Assessment in Probation Services: An Evaluation*, Research Study 211, London: Home Office.

Redondo, S., Sanchez-Meca, J. and Garrido, V. (2002) 'Crime treatment in Europe: A review of outcome studies', in J. McGuire (ed.) *Offender Rehabilitation and Treatment*, Chichester: Wiley.

Reichman, N. (1986) 'Managing crime risks: Toward an insurance based model of social control', *Research in Law, Deviance and Social Control*, 8, 151–72.

Reid, W.J. and Epstein, L. (1972) *Task Centred Casework*, New York: Columbia University Press.

Reid, W.J. and Shyne, A. (1968) *Brief and Extended Casework*, New York: Columbia University Press.

Research, Development and Statistics (2004a) *Home Office RDS and YJB Minimum Standards for Reconviction Studies*, internal paper.

Research, Development and Statistics (2004b) *Home Office RDS and YJB Standards for Impact Studies in Correctional Settings*, internal paper.

Rex, S. and Matravers, A. (eds) (1998) *Pro-Social Modelling and Legitimacy*, Cambridge: Institute of Criminology.

Rex, S. (1999) 'Desistance from offending: Experiences of probation', *Howard Journal of Criminal Justice*, 38, 4, 366–83.

Rex, S. and Gelsthorpe, L. (2002) 'The role of community service in reducing offending: Evaluating pathfinder projects in the UK', *Howard Journal*, 41, 311–25.

Rex, S. and Matravers, A. (eds) (1998) *Pro-Social Modelling and Legitimacy*, Cambridge: Institute of Criminology.

Rex, S., Gelsthorpe, L., Roberts, C. and Jordan, P. (2004) *What's Promising in Community Service: Implementation of Seven Pathfinder Projects*, Home Office Research Findings 231, London: Home Office.

Rex, S., Lieb, R., Bottoms, A. and Wilson, L. (2003) *Accrediting Offender Programmes: A Process-based Evaluation of the Joint Prison/Probation Services Accreditation Panel*, Home Office Research Study 273, London: Home Office.

Richie, J.H., Dick, D. and Lingham, R. (1994) *The Report into the Care and Treatment of Christopher Clunis*, London: HMSO.

Richmond, M. (1917) *Social Diagnosis*, New York: Russell Sage Foundation.

Roberton, A. (1961) 'Casework in Borstal', *Prison Service Journal*, 1, 15–22.

Roberts, C. (1989) *Hereford and Worcester Probation Service Young Offender Project: First Evaluation Report*, Oxford: Department of Social and Administrative Studies.

Roberts, C. (2004) 'An early evaluation of a cognitive offending behaviour programme ("Think First") in probation areas', *VISTA*, 8, 137–45.

Roberts, R. and Nee, R. (1970) *Theories of Social Casework*, Chicago: University of Chicago Press.

Robinson, D. (1995) *The Impact of Cognitive Skills Training on Post-release Recidivism among Canadian Federal Offenders*, No. R-41 Research Branch, Ottawa: Correctional Service Canada.

Robinson, G. (1999) 'Risk management and rehabilitation in the probation service: Collision and collusion', *The Howard Journal*, 38, 4, 421–33.

Robinson, G. (2001) 'Power, knowledge and What Works in probation', *Howard Journal*, 40, 235–54.

Robinson, G. (2002) 'Exploring risk management in probation practice: Contemporary developments in England and Wales', *Punishment and Society*, 4, 5–25.

Robinson, G. (2005) 'What works in offender management?', *Howard Journal of Criminal Justice*, 44, 3, 307–18.

Robinson, G. and Dignan, J. (2004) 'Sentence management', in A. Bottoms, S. Rex and G. Robinson (eds) *Alternatives to Prison: Options for an Insecure Society*, Cullompton: Willan.

Robinson, V. (ed.) (1962) *Jessie Taft, Therapist and Social Work Educator*, Philadelphia: University of Philadelphia Press.

Rogers, C.R. (1951) *Client-Centred Therapy: Its Current Practice, Implications and Theory*, London: Constable.

Rorty, R. (1993) 'Human rights, rationality and sentimentality', in S. Shute and S. Hurley (eds) *On Human Rights: The Oxford Amnesty Lectures 1993*, New York: Basic Books.

Rose, N. (1985) *The Psychological Complex: Power, Politics and Society in England 1869–1939*, London: Routledge.

Rose, N. (2000) 'Government and control', *British Journal of Criminology*, 40, 321–39.

Rosenhan, D.L. (1973) 'On being sane in insane places', *Science* 19 January 1973, reprinted in D. Krebs (ed.) *Readings in Social Psychology*, New York: Harper and Row, 1982, 264–74.

Ross, R.R. and Fabiano, E.A. (1985) *Time to Think: A Cognitive Model of Delinquency Prevention and Offender Rehabilitation*, Johnson City, Tennessee: Institute of Social Sciences and Arts.

Ross, R.R., Fabiano, E.A. and Ewles, C.D. (1988) 'Reasoning and rehabilitation', *International Journal of Offender Therapy and Comparative Criminology*, 32, 29–35.

Ross, R.R., Fabiano, E.A. and Ross, R.D. (1986) *Reasoning and Rehabilitation: A Handbook for Teaching Cognitive Skills*, Ottawa: University of Ottawa.

Rotman, E. (1990) *Beyond Punishment: A New View of the Rehabilitation of Offenders*, Westport, Conn.: Greenwood Press.

Rutherford, A. (1994) *Criminal Justice and the Pursuit of Decency*, Winchester: Waterside Press.

Rutherford, A. (1986) *Growing Out of Crime*, Harmondsworth: Penguin.

Rutherford, A. (1998) 'Criminal policy and the eliminative ideal', in C. Jones Finer and M. Nellis (eds) *Crime and Social Exclusion*, Oxford: Blackwell.

Salzberger-Wittenberg, I. (1970) *Psycho-Analytic Insight and Relationships: A Kleinian Approach*, London: Routledge.

Sampson, R.J. and Laub, J.H. (1993) *Crime in the Making: Pathways and Turning Points Through Life*, London: Harvard University Press.

Scheff, T.J. (1966) *Being Mentally Ill*, Chicago: Aldine.

Schiff, M. (2003) 'Models, challenges and the promise of restorative conferencing strategies', in A. von Hirsch, J. Roberts, A.E. Bottoms, K. Roach and M. Schiff (eds) *Restorative Justice and Criminal Justice: Competing or Reconcilable Paradigms?* Oxford: Hart Publishing.

Schur, E. (1973) *Radical Non-Intervention*, New Jersey: Prentice-Hall.

Scottish Executive (2004) *Scotland's Criminal Justice Plan*, Edinburgh: Scottish Executive.

Seddon, M. (1979) 'Clients as Social Workers', in D. Brandon and B. Jordan (eds) *Creative Social Work*, Oxford: Blackwell.

Seebohm, F. (Chairman) (1968) *Report of the Committee on Local Authority and Allied Personal Social Services*, Cmnd 3703, London: HMSO.

Shapland, J., Atkinson, A., Colledge, E., Dignan, J., Howes, M., Johnstone, J., Pennant, R., Robinson, G. and Sorsby, A. (2004) *Implementing Restorative Justice Schemes (Crime Reduction Programme): A Report on the First Year*, Home Office Online Report 32/04.

Shapland, J., Atkinson, A., Colledge, E., Dignan, J., Howes, M., Johnstone, J., Pennant, R., Robinson, G. and Sorsby, A. (2002) 'Evaluating the fit: Restorative justice and criminal justice', paper to the British Criminology Conference, unpublished.

References 197

Shaw, M. (1974) *Social Work in Prisons*, Home Office Research Study 22, London: HMSO.

Shaw, M. and Hannah-Moffatt, K. (2000). 'Gender, diversity and risk assessment in Canadian corrections', *Probation Journal*, 47, 163–72.

Shaw, M. and Hannah-Moffatt, K. (2004) 'How cognitive skills forgot about gender and diversity', in G. Mair (ed.) *What Matters In Probation*, Cullompton: Willan.

Shearing, C.D. and Stenning, P.C. (1985) 'From the panopticon to Disney World: The development of discipline', in A.N. Doob and E.L. Greenspan (eds) *Perspectives in Criminal Law*, Ontario: Canada Law Book Inc.

Sherman, L., Strang, H. and Woods, D.J. (2000) *Recidivism Patterns in the Canberra Reintegrative Shaming Experiments (RISE)*. Available at: www.aic.gov.au/rjustice/rise/recidivism/.

Sherman, L., Farrington, D., Welsh, B. and Mackenzie D.L. (eds) (2002) *Evidence-Based Crime Prevention*, New York: Routledge.

Sherman, L., Gottfredson, D., MacKenzie, D., Eck, J., Reuter, P. and Bushway, S. (1998) *Preventing Crime: What Works, What Doesn't, What's Promising*, Washington: National Institute of Justice.

Shute, S. (2003) 'The development of parole and the role of research in its reform', in L. Zedner and A. Ashworth (eds) *The Criminological Foundations of Penal Policy: Essays in Honour of Roger Hood*, Oxford: Oxford University Press.

Sillitoe, A. (1959) *The Loneliness of the Long-distance Runner*, London: W.H. Allen.

Simey, T. (1957) 'Social research and social casework I', in E.M. Goldberg, E. Irvine, A. Lloyd Davies and K. McDougall (eds) *The Boundaries of Casework*, London: Association of Psychiatric Social Workers.

Simon, J. (1987) 'The emergence of a risk society: Insurance, law and the state', *Socialist Review*, 95, 61–89.

Simon, J. (1988) 'The ideological effects of actuarial practices', *Law & Society Review*, 22, 771–800.

Simon, J. (1993) *Poor Discipline: Parole and the Social Control of the Underclass 1890–1990*, London: University of Chicago Press.

Simpkin, M. (1979) *Trapped within Welfare*, London: Macmillan.

Sinclair, I. (1971) *Hostels for Probationers*, Home Office Research Study 6, London: HMSO.

Sinfield, A. (1969) *Which Way for Social Work?*, London: Fabian Society.

Smith, D. (1995) *Criminology for Social Work*, Basingstoke: Macmillan.

Smith, G. and Cantley, C. (1984) 'Pluralistic evaluation', in J. Lishman (ed.) *Evaluation*, Aberdeen: University of Aberdeen.

Social Exclusion Unit (2002) *Reducing Re-offending by Ex-prisoners*, London: Office of the Deputy Prime Minister.

Spencer, J. (1995) 'A response to Mike Nellis: Probation values for the 1990s', *Howard Journal*, 34, 3, 344–49.

Stephens, K., Coombs, J. and Debidin, M. (2004) *Black and Asian Offenders Pathfinder: Implementation Report*, Development and Practice Report 24, London: Home Office.

Strawson, P.F. (1968) 'Freedom and resentment', in Strawson (ed.) *Studies in the Philosophy of Thought and Action*, Oxford: Oxford University Press.

Sugg, D. (2000) *Aggression Replacement Training: Research Report*, report to Correctional Services Accreditation Panel, unpublished.

Sullivan, M. (1996) *The Development of the British Welfare State*, London: Prentice-Hall.

Sykes, G.M. and Matza, D. (1957) 'Techniques of neutralization: A theory of delinquency', *American Sociological Review*, 22, 664–73.

Taft, J. (1913) 'The Woman Movement from the Point of View of Social Consciousness', dissertation submitted to the Department of Philosophy, University of Chicago.

Taft, J. (1926) 'The relationship of psychiatry to social work' (paper to the New York City Conference of Charities and Corrections May 12, 1926), reprinted in V. Robinson (ed.) *Jessie Taft, Therapist and Social Work Educator*, Philadelphia: University of Philadelphia Press, 1962, pp. 56–63.

Taft, J. (1958) *Otto Rank, a Biographical Study Based on Notebooks, Letters, Collected Writings, Therapeutic Achievement and Personal Associations*, New York: The Julian Press.

Taylor, L. and Taylor, I. (1968) 'We are all deviants now – some comments on crime', *International Socialism*, 34, 29–32.

Thornton, D. (1987) 'Treatment effects on recidivism: A reappraisal of the nothing works doctrine', in B. McGurk, D. Thornton and M. Williams (eds) *Applying Psychology to Imprisonment: Theory and Practice*, London: HMSO.

Thorpe, D.H., Smith, D., Green, C.J. and Paley, J. (1980) *Out of Care*, London: Allen and Unwin.

Timms, N. (1962) *Casework in the Child Care Service*, London: Butterworths.

Toch, H. (2000) 'Altruistic activity as correctional treatment', *International Journal of Offender Therapy and Comparative Criminology*, 44, 270–78.

Towle, C. (1954) *The Learner in Education for the Professions*, Chicago: University of Chicago Press.

Travis, A. (1995) 'Probation officers defend training', *The Guardian*, 23 February, 6.

Travis, J. and Petersilia, J. (2001) 'Reentry reconsidered: A new look at an old question', *Crime and Delinquency*, 47, 3, 291–313.

Trotter, C. (1993) *The Supervision of Offenders – What Works? A Study Undertaken in Community Based Corrections, Victoria*, Melbourne: Social Work Department, Monash University and Victoria Department of Justice.

Trotter, C. (1996) 'The impact of different supervision practices in community corrections', *Australian and New Zealand Journal of Criminology*, 28, 2, 29–46.

Trotter, C. (1999) *Working with Involuntary Clients: A Guide to Practice*, London: Sage.

Trotter, C. (2001) *Focus on People: Effect Change*, Dinas Powys: Cognitive Centre Foundation.

Truax, C. and Carkhuff, R. (1967) *Towards Effective Counselling and Psychotherapy*, Chicago: Aldine.

Turnbull, P., McSweeney, T., Webster, R., Edmunds, M. and Hough, M. (2000) *Drug Treatment and Testing Orders: Final Evaluation Report*, Home Office Research Study 212. London: Home Office.

Underdown, A. (1998) *Strategies for Effective Supervision: Report of the HMIP What Works Project*, London: Home Office.

Urmson, J. (1953) 'The interpretation of the moral philosophy of J. S. Mill', *Philosophical Quarterly*, 3, 33–39.

Van Ness, D. and Strong, K.H. (1997) *Restoring Justice*, Cincinnati, Ohio: Anderson Publishing.

Vanstone, M. (1993) 'A 'missed opportunity' re-assessed: The influence of the day training centre experiment on the criminal justice system and probation policy and practice', *British Journal of Social Work*, 23, 213–29.

Vanstone, M. (2000) 'Cognitive-behavioural work with offenders in the UK: A history of influential endeavour', *The Howard Journal*, 39, 2, 171–83.

Vanstone, M. (2004) *Supervising Offenders in the Community: A History of Probation Theory and Practice*, Aldershot: Ashgate.

von Hirsch, A. (1976) *Doing Justice: The Choice of Punishments*, Report of the Committee for the Study of Incarceration, New York: Hill and Wang.

von Hirsch, A. (1986) *Past or Future Crimes*, Manchester: Manchester University Press.

von Hirsch, A. and Maher, L. (1992) 'Should penal rehabilitationism be revived?', *Criminal Justice Ethics*, Winter/Spring, 25–30.

von Hirsch, A., Roberts, J., Bottoms, A.E., Roach, K. and Schiff, M. (eds) (2003) *Restorative Justice and Criminal Justice: Competing or Reconcilable Paradigms?* Oxford: Hart Publishing.

Walgrave, L. (2001) 'On restoration and punishment: Favourable similarities and fortunate differences', in A. Morris and G. Maxwell (eds) *Restorative Justice for Juveniles: Conferencing, Mediation and Circles*, Oxford: Hart.

Walker, H. and Beaumont, W. (1981) *Probation Work: Critical Theory and Socialist Practice*, Oxford: Blackwell.

Walker, N., Farrington, D. and Tucker, G. (1981) 'Reconviction rates of adult males after different sentences', *British Journal of Criminology*, 21, 357–60.

Wallace, D. (1967) 'The Chemung County evaluation of casework service to dependent multiproblem families: Another problem outcome', *Social Service Review*, 41, 379–89.

Wasik, M. and von Hirsch, A. (1988) 'Non-custodial penalties and the principles of desert', *Criminal Law Review*, 555–571.

Wilcox, A. (2003) 'Evidence-based youth justice? Some valuable lessons from an evaluation for the Youth Justice Board', *Youth Justice*, 3, 19–33.

Wilcox, A., Young, R. and Hoyle, C. (2004) *Two-Year Resanctioning Study: A Comparison of Restorative and Traditional Cautions*, Home Office Online Report 57/04.

Wilkins, L.T. (1958) 'A small comparative study of the results of probation', *British Journal of Delinquency*, 8, 201–209.

Wilkins, L.T. (1964) *Social Deviance*, London: Tavistock.

Wilkinson, J. (1997) 'The impact of Ilderton Motor Project on motor vehicle crime and offending', *British Journal of Criminology*, 37, 568–81.

Wilkinson, J. (1998) *Developing the evidence-base for probation programmes*, PhD thesis, University of Surrey.

Willis, A. (1983) 'The balance between care and control in probation: A research note', *British Journal of Social Work*, 13, 339–46.

Winnicott, C. (1962) 'Casework and agency function', *Case Conference*, 8, 178–84.

Wodarski, J. and Bagarozzi, D. (1979) 'A review of the empirical status of traditional modes of interpersonal helping: Implications for social work practice', *Clinical Social Work*, 7, 231–55.

Woolf, Lord Justice (1991) *Prison Disturbances April 1990: Report of an Inquiry*, Cm. 1456, London: HMSO.

Wootton, B. (1959) *Social Science and Social Pathology*, London: Allen and Unwin.

Wright, M. (1991) *Justice for Victims and Offenders: A Restorative Response to Crime*, Milton Keynes: Open University Press.

Young, J. (1971) *The Drugtakers*, London: Paladin.

Young, P. (1967) *The Student and Supervision in Social Work Education*, London: Routledge.

Zamble, E. and Quinsey, V. (1997) *The Criminal Recidivism Process*, Cambridge: Cambridge University Press.

Zedner, L. (1995) 'Wayward sisters: The prison for women', in N. Morris and D.J. Rothman (eds) *The Oxford History of the Prison*, Oxford: Oxford University Press.

Zehr, H. (1990) *Changing Lenses: A New Focus for Crime and Justice*, Scottdale, PA: Herald Press.

Index

accreditation, *see* Correctional Services Accreditation Panel (CSAP)
Acts of Parliament
  Children and Young Persons Act (1969), 71, 83
  Criminal Justice Act (1967), 81
  Criminal Justice Act (1982), 83, 101
  Criminal Justice Act (1991), 85, 89, 90, 101–2, 105
  Criminal Justice Act (2003), 152, 161, 176
  Penitentiary Act (1779), 37–8, 41
  Prevention of Crime Act (1908), 49–50
  Probation of Offenders Act (1907), 51
  Rehabilitation of Offenders Act (1974), 9, 13, 146
  Summary Jurisdiction Act (1879), 45
aftercare, *see* resettlement
alcohol, *see* substance misuse
Allen, F., 1, 2
American Friends Service Committee, 26, 72, 84
Asian offenders, 126, 171

Bazemore, G., 138, 141–3, 155
Beccaria, C., 20, 36, 84, 173
Bentham, J., 20, 36–8, 173
bifurcation, 84, 94
Black offenders, 126
Booth, W., 18–19, 31
Borstal institutions, 25, 48, 50–1, 58, 162
Bottoms, A., 1, 7, 25, 80, 81–2, 83, 84, 85, 94–6, 103, 148, 175
Braithwaite, J., 29, 136, 137, 139–41, 148, 167
Burt, C., 51

Cambridge-Somerville Youth Study, 59–61
capital punishment, 23, 34

Carnarvon Committee, 39
Carpenter, M., 44
case management, 90, 116, 119, 125–6, 127, 128, 129, 158
  see also supervisory relationship
charity, 18, 42–3, 52–3, 75
Charity Organisation Society (COS), 52, 54
Church of England Temperance Society (CETS), 17–18, 45
  see also religion
citizenship, 10–12, 140
cognitive-behavioural approaches, 105–6, 108, 111–12, 113, 118, 119–20, 121, 126, 140, 147, 153, 163
Cohen, S., 12, 69, 70, 71, 88, 93, 102–3, 166
combination orders, 102, 176
communities, 28–9, 31, 135, 137, 140–2, 149, 150–1, 153–4, 155–7, 166–9, 177
community justice, 167
community punishment, *see* community service
community service, 80, 81, 85, 94, 97, 100, 102, 103, 113, 117, 129, 142, 158, 175
Correctional Services Accreditation Panel (CSAP), 108, 117, 120, 125, 128
Cox, E., 45
Crime Reduction Programme (CRP), 1, 113, 120, 122, 125, 126, 127–8, 137, 159, 162, 176
critical criminology, 68–70, 132

desert, *see* 'just deserts'
desistance, 5, 12–13, 15, 134, 143–8, 154–8, 165–6, 170, 177
  ontogenic explanations, 144
  sociogenic explanations, 144

Detention Centres, 50
'differentiation of discipline' thesis, 94–5, 96
discharged prisoners' aid societies (DPAS), 43, 81, 161
discipline, *see* 'differentiation of discipline' thesis; Foucault, M.; 'penitentiary discipline'
drugs, *see* substance misuse
Du Cane, E., 40, 174

employment, 8, 10, 41, 42, 43, 44, 45, 69, 113, 124, 144, 145, 147, 149–51, 152, 157, 166
see also prison labour
eugenics, 51, 53
see also heredity

Feeley, M. and Simon, J., 1, 88–9, 92, 94, 96, 132, 160
female offenders, 27, 42, 44, 53, 54, 60–1, 67, 126, 176
Fielding, H., 34–5, 173
Foucault, M., 5, 32, 86, 88
Freagarrach Project, 119

Garland, D., 5, 6, 7, 9, 19, 28, 47, 48, 73, 76, 80, 85, 87, 89, 95–6, 159, 160, 171
Gladstone, H., 14, 33, 40, 46–8, 49, 174

Her Majesty's Inspectorate of Probation (HMIP), 90–1, 112, 113, 125, 150
heredity, 18, 39
see also eugenics
Holford Committee, 38
hostels, 43, 68, 100
houses of correction, 33–5, 38, 43
Howard, J., 34–7, 47, 173
Hudson, B., 2, 4, 6, 7, 22, 26, 27, 86, 106, 169, 172
human capital, 147

IMPACT study, 66–7, 100, 175–6
indeterminate sentencing, 20–1, 72, 84, 159
individualization, 7, 20–1, 47

Intensive Supervision and Surveillance Programme (ISSP), 118–19
Intermediate Treatment (IT), 83, 95, 101

'just deserts', 25–7, 72–3, 74, 84–5, 89
juvenile offenders, *see* young offenders

Kant, I., 25, 27

labelling theory, 68–70
'less eligibility', 43, 44

McGuire, J., 28, 93, 104–5, 106, 108, 109, 113, 121, 128, 129, 168
McWilliams, W., 9, 17, 25, 27, 45, 80, 81–2, 86, 148, 170
Mannheim, H., 9, 20–1, 22, 56
Martinson, R., 65–6, 67, 99–100
Maruna, S., 29, 143–6, 148, 149, 153, 156, 166, 177
May Committee, 79–80
media, 40, 41
'mental hygiene' movement, 52–4
meta-analysis, 106–10, 121, 125
Missionaries, *see* religion

National Association of Probation Officers (NAPO), 65, 93, 125, 132
National Offender Management Service (NOMS), 152, 158, 160–1, 166
National Probation Service (NPS), 114, 124
see also probation
'New Careers' movement, 167
New Labour, 105
'new penology' thesis, *see* Feeley, M. and Simon, J.
non-treatment paradigm, 15, 81–2, 94, 134, 148, 154
'Nothing works', 15, 49, 64–70, 73–5, 93, 98–101, 104, 110, 131, 159

Panopticon, 36–8
'Pathfinders', 113, 115–20, 124–8, 131, 162–5
see also Crime Reduction Programme (CRP); programmes

penal servitude, 40, 43, 174
penitentiary, *see* prisons
'penitentiary discipline', 34–7
Poor Law, 18
'populist punitiveness', 85, 159, 175
poverty, 18, 27, 29, 33, 34, 43, 52, 64,
    73, 75–6, 105, 169
'prison works', 85, 104
prisons, 8, 13, 14, 29, 32, 34–41, 44,
    46–8, 65, 72, 79–80, 85, 88–9, 95,
    100, 101–5, 113, 114, 130–1,
    160–1, 166, 170–1, 174,
    177–8
  labour, 33–4, 36–9, 43, 47,
    50, 173
  Millbank, 39
  monastic, 33
  Parkhurst, 44
  Pentonville, 32, 39, 173
  'silent system', 39
pro-social modelling, 117, 129
probation, 17, 21, 22, 24–5, 44–6,
    48, 51–2, 55–9, 64–9, 80–6,
    89–95, 98–132, 146–8,
    150, 157, 160–5, 166,
    168–9, 170, 174, 177
Probation of First Offenders Bill, 46
programmes, 28, 94, 95, 101, 106–13,
    115–16, 118–22, 124–31,
    141–2, 150
  Aggression Replacement Therapy,
    120
  definitions, 128
  Enhanced Thinking Skills, 115
  Priestley, 1, 115
  Reasoning and Rehabilitation, 106,
    111, 113, 115
  Straight Thinking on Probation,
    111–12, 113, 130–1, 176
  Think First, 115–16
  see also cognitive-behavioural
    approaches
psychiatry, 20, 24, 51, 55–6, 58, 63, 69
psychology, 20, 28, 51–6, 63–4, 69, 75,
    76, 88, 94, 104, 105–6, 138
  materialist, 36–7
public protection, 15, 28, 84, 85, 90–1,
    95–6, 114
punishment in the community, 85

Radzinowicz, L., 6, 22, 24, 40, 42,
    43, 44, 45, 48, 49, 50, 56,
    57, 66, 174
reentry, *see* reentry court; resettlement
reentry court, 153–4, 168
Reformatory system, 43–4, 47
  Elmira, 49–50
rehabilitation
  correctional model, 5–7, 12, 13,
    30, 86, 96–7, 134, 140, 154,
    156, 157–8
  definitions, 1–14, 96–7, 166
  devolution of, 141, 156, 157
  justifications for, 14, 16–31
  legal context, 9–11, 12, 13
  medical context, 2–3, 8
  'naturalistic', *see* desistance
  'personalist' approach, 27
  and reform, 1, 4, 6–7, 175
  reintegration model, 7–13, 97, 138,
    139, 140–2, 146, 148–9,
    151–4, 155–8, 167
  'relational', 155–6, 157–8
  'state-obligated', 27, 29, 86, 151, 169
  strength-based approaches, 29, 141,
    166–8
reintegrative shaming, *see* restorative
    justice
religion, 17–19, 24, 26, 30, 34–5, 37,
    38, 39, 40, 45, 47, 50
resettlement, 7–8, 11, 18–19, 26, 41–3,
    50, 80–1, 97, 100, 101–2, 114,
    117, 124, 134, 148–58, 161–6,
    168, 173
'responsibilization', 73, 74, 156, 167
restorative justice, 15, 29, 134–43, 145,
    146, 148, 154–8, 167, 176
Richmond, M., 52–3
rights of offenders, 17, 23–7, 30
risk, 30, 87, 90, 122
  assessment of, 21, 28, 89, 90–2,
    109, 118, 122–4, 126, 160
  management of, 86–92, 132
Ross, R., 67–8, 100, 105, 106, 111, 112

Scotland, 59, 112, 119, 147,
    153, 160, 178
Seebohm Committee, 58, 75
sex offenders, 119–20, 168, 177

shaming, *see* restorative justice
social capital, 147–8, 154, 155,
    156, 157
social casework, 51–64, 75–7
    task-centred, 62, 94, 129
social contract theory, 27, 169–70
social exclusion, 139, 150, 157
Social Exclusion Unit (SEU), 149,
    150, 152, 162
social work, 15, 20, 24–5, 30, 49,
    51–65, 69–77, 81, 85, 98,
    128, 147–8, 175
    'radical', 64, 65, 70, 74
    training, 54–5, 58, 62–3, 65,
        75, 77, 128
Statement of National Objectives and
    Priorities (SNOP), 83–4,
    157, 162
substance misuse, 3, 17–18, 45, 60,
    115, 118, 129, 148, 152, 166,
    167, 177
supervisory relationship, 25, 57–8,
    60–2, 82, 90, 128–30, 142–3,
    146–8, 157–8
    see also case management

Taft, J., 52–4
therapeutic jurisprudence, 168
therapeutic relationship, *see*
    supervisory relationship
'ticket of leave' system, 41–2, 46, 174
transportation, 34, 38, 40, 41, 42, 44,
    173, 174

treatment model, 5, 15, 23–5, 28, 30,
    55, 70–3, 78, 81, 93–4, 136,
    137–8, 143

Utilitarianism, 20–6, 28, 170

Vanstone, M., 17, 28, 44, 45, 46, 52,
    59, 77, 93, 94, 104, 110, 111, 125,
    129, 130, 148
victims of crime, 28, 30, 86, 122,
    134, 135–41, 149, 155,
    156, 176
Vincent, H., 45–6, 174
'Vocational High' experiment, 60–1
voluntarism, 23–5, 57–8, 72
von Hirsch, A., 6, 14, 26, 72,
    84, 136

Welfare State, 19, 22, 27, 30, 49, 73–6
'what works', 15, 16, 98–133, 177
    conferences, 112
    critiques, 132
women, *see* female offenders

young offenders, 39, 43–4, 47, 49–51,
    59–61, 71, 82–3, 93–4, 95,
    101, 102, 103, 108, 113,
    118–19, 130, 138, 141,
    144, 162, 175
Youth Inclusion Programme, 119
Youth Justice Board, 113, 118
Youth Offending Teams (YOTs),
    113, 114